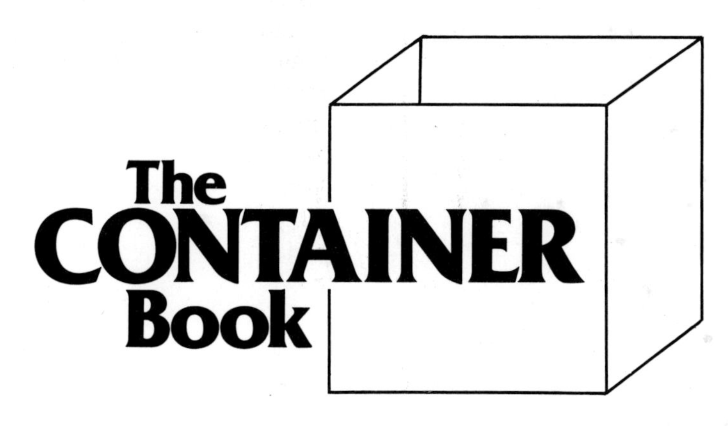
The
CONTAINER
Book

Other Books by the Authors

Thelma R. Newman

Creative Candlemaking
Contemporary Decoupage
Contemporary African Arts and Crafts
Contemporary Southeast Asian Arts and Crafts
The Complete Book of Making Miniatures
Quilting, Patchwork, Appliqué, and Trapunto
Leather as Art and Craft
Crafting with Plastics
Plastics as an Art Form
Plastics as Design Form
Plastics as Sculpture
Woodcraft
Wax as Art Form

With Jay Hartley Newman and Lee Scott Newman

The Frame Book
The Lamp and Lighting Book
Paper as Art and Craft

Jay Hartley Newman and Lee Scott Newman

Plastics for the Craftsman
Kitecraft
Wire Art

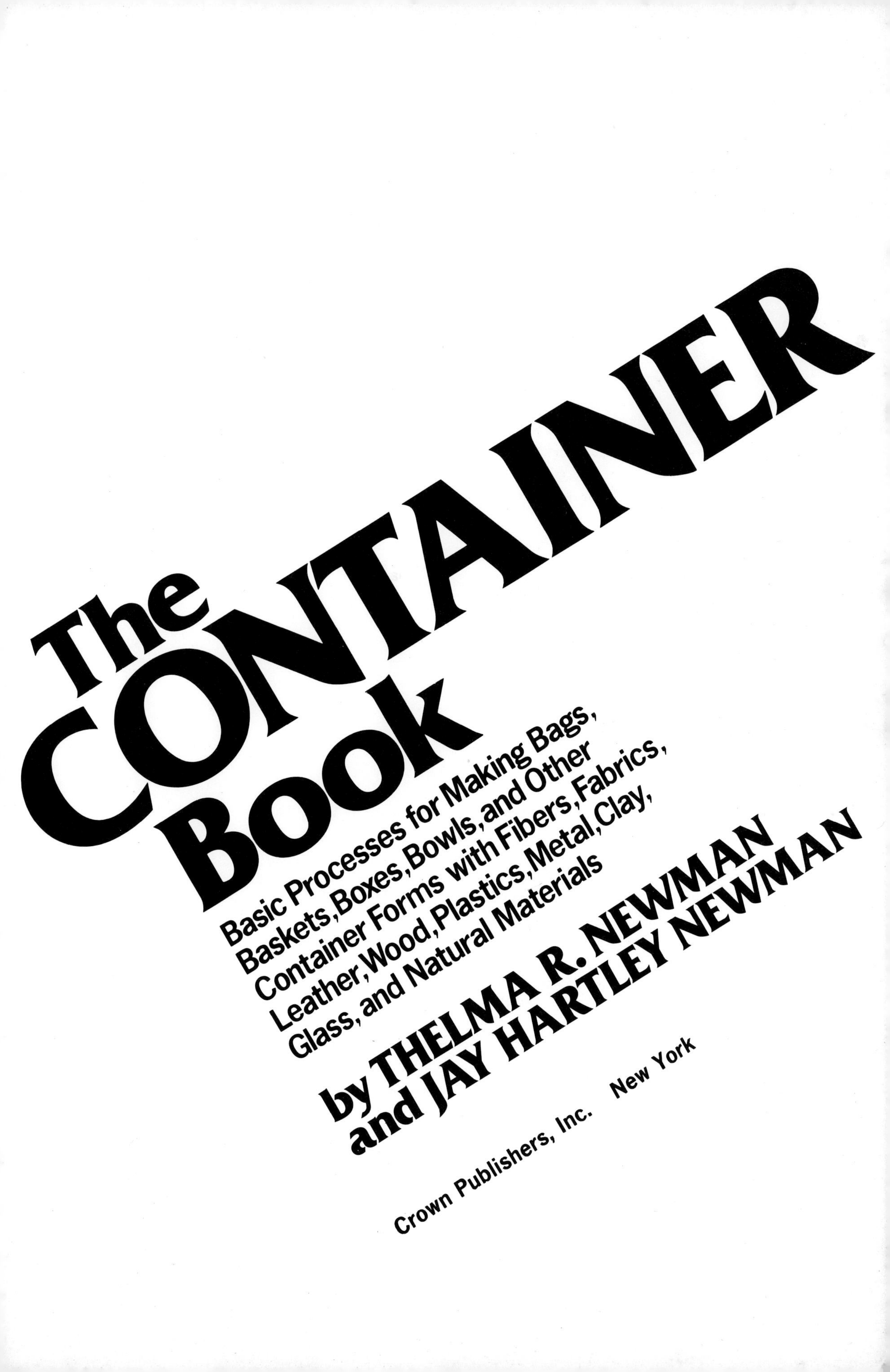

The CONTAINER Book

Basic Processes for Making Bags, Baskets, Boxes, Bowls, and Other Container Forms with Fibers, Fabrics, Leather, Wood, Plastics, Metal, Clay, Glass, and Natural Materials

by THELMA R. NEWMAN
and JAY HARTLEY NEWMAN

Crown Publishers, Inc. New York

For Connie Joy

All photographs by the authors unless otherwise noted.

Designed by Laurie Zuckerman

Inquiries should be addressed to Crown Publishers, Inc.,
One Park Avenue, New York, N.Y. 10016.

Printed in the United States of America
Published simultaneously in Canada by
General Publishing Company Limited

LIBRARY OF CONGRESS CATALOGING IN PUBLICATION DATA

Newman, Thelma R
The container book.

Includes index.
1. Handicraft. 2. Containers. I. Newman,
Jay Hartley, joint author. II. Title.
TT157.N43 1977 745.59'3 77-401
ISBN 0-517-52897-5
ISBN 0-517-52898-3 pbk.

CONTENTS

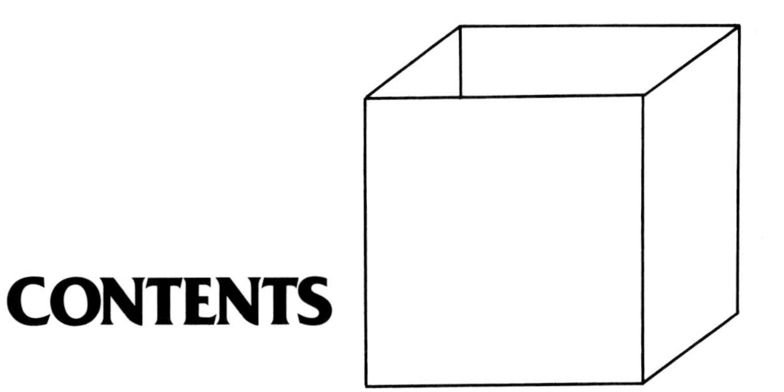

ACKNOWLEDGMENTS

More than anything else, the help and encouragement of the many contributors to this book made it what it is; to them goes our deep appreciation.

In addition, we owe a great debt of gratitude to the individuals who went to extraordinary lengths for this book. Timothy M. Glotzbach, William O. Huggins, and Joseph S. Clift, students of L. Brent Kington at Southern Illinois University, wrote most of the section on electroforming and electroplating, and Joseph S. Clift executed an electroformed container especially to show the steps involved in that process. David Luck, a student of Julius Schmidt and Chunghi Choo at the University of Iowa, illustrated and captioned iron casting and electroforming processes as they were carried out at the University of Iowa Graduate Art Department. Jan Axel meticulously detailed the process of creating one of her ceramic containers, and Robert Axel, her husband, documented the steps. We thank them all for their effort, time, interest, and skill.

We owe very special thanks as well to a group of artists and craftsmen who gave us enormous help by allowing us to photograph them at work, arranging to have photographic sequences taken, and supplying photographs of their work: Renie Breskin Adams, Carol Austin, Dave Boronda, Harlan Butt, Art (Espenet) Carpenter, Sas Colby, Julie Connell, Anna Continos, Don Drumm, Sally Edmondson, Malcolm Fatzer, Susan Felix, Verne Funk, Marc Goldring, Maud Guilfoyle, Jan Hoffman, William O. Huggins, Susan Jamart, L. Brent Kington, Mary Kretsinger, Marcia Lewis, Mark Lindquist, Melvin Lindquist, Marcia Lloyd, David Luck, Dick Luster, Ralph Massey, Sylvia Massey, Joel Moses, William Rockhill Nelson Gallery of Art, Barbara Anne Nilhaūsen, Yvonne Porcella, Wayne Raab, Rohm & Haas Company, Albert Rosenblatt, Ed Rossbach, Bob Stocksdale, Joy Stocksdale, Marian Slepian, Paul Taylor, Lucy Traber, Kay Wiener.

As ever, we are grateful to Norm Smith for fine photo processing, and our thanks go to Pat Weidner, Gal Saturday, as well.

Last, but never least, our deepest thanks to Jack Newman, husband and father, and Lee Newman, son and brother, for their unending and essential advice, support, and encouragement.

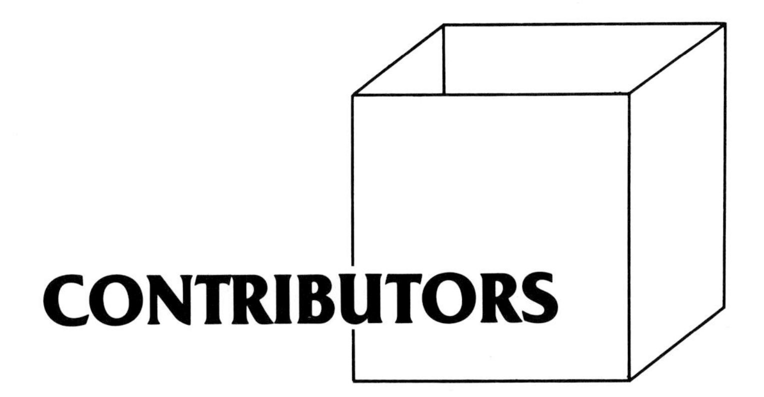

CONTRIBUTORS

Renie Breskin Adams
Judith Anderson
Arno Foil Tape
Atkins Gallery of Fine Arts
Carole Austin
Jan Axel
Alice Balterman
Thomas R. Bambas
Suzanne Benton
Scott Bliss
Dave Boronda
Harlan Butt
Erni Cabat
Rose Cabat
Jim Cantrell
Art (Espenet) Carpenter
Shirley Charron
Joseph S. Clift
Florence Cohen
Sas Colby
Julie Connell
Container Corporation of America
Anna Continos
Jim Cotter
Danese, Milano
Elisa D'Arrigo
Patricia J. Daunis-Dunning
Dee Davis
Richard R. Dehr
Paul A. Diekmeyer
Don Drumm
Lisa Drumm
Walter Dyer
Sally Edmondson
Greer Farris
Malcolm Fatzer
Jeanette Feldman
Susan Felix
Fred Fenster
Fibre Box Association
Froelich Leather Co.
Verne Funk
Angelo C. Garzio
Imogene Gieling
Timothy M. Glotzbach
Marc Goldring
Lida Gordon
Grainware Company
Albert Green
Gary Groves (Woodworks)
Barbara Grygutis
Maud Guilfoyle
Elizabeth Gurrier
Jim Hanko
Eileen Gilbert Hill
Jan Hoffmann
William O. Huggins
Sylvia Hyman
Kent F. Ipsen
Marsha Anne Isoshima

Susan Jamart
Japan National Tourist Organization
Eduardo Ayala Jimenez
Peggy Kent
Daniella Kerner
L. Brent Kington
Vivian Kline
Earl Krentzin
Mary Kretsinger
Kruger Van Eerde Gallery
Murry Kusmin
Margaret Laws
Marcia Lewis
Mark Lindquist
Melvin Lindquist
Marcia Lloyd
Jan Brooks Loyd
David Luck
Dick Luster
Kathryn McBride
Jean Mann
Enzo Mari
Ralph Massey
Sylvia Massey
Richard Mawdsley
Metropolitan Museum of Art
Stephen Miller
Norma Minkowitz
Barbara Minor
Steve Mirones
Joel Moses
Marcia Moss
Joyce Moty
Museum of Contemporary Crafts
Joy Nagy
William Rockhill Nelson Gallery of Art
Barbara Anne Nilaūsen
Olinkraft, Inc.
Raymond Pelton
Nancy M. Piatkowski
Judith Plotner
Stanley Plotner
Yvonne Porcella
Wayne Raab
Ruth Rickard
Michael B. Riegel
Louise Robbins
Rohm & Haas Company
Albert Rosenblatt
Hal Ross
Ed Rossbach
Judith Rothbart
Ginna Sadler
Carol Savid
Herb Schumacher
June Schwarcz
Kay Sekimachi
Helen Shirk
Thomas Charles Siefke
Jeff Slaboden
Marian Slepian
Neal Small
Joan Sterrenburg
Rebecca A. T. Stevens
Marguerite Stix
Bob Stocksdale
Joy Stocksdale
Ann Swan
Paul Taylor
John Teeble
Thermoplastic Processes, Inc.
Lucy Traber
Oppi Untracht
Saara Hopea-Untracht
George P. van Duinwyk
Weyerhaeuser Co.
Kay Wiener
Andrew Willner
Diana Schmidt Willner
Paula Winokur
Robert M. Winokur
Fred J. Woell
Chris Wright
Frank Wright

PREFACE

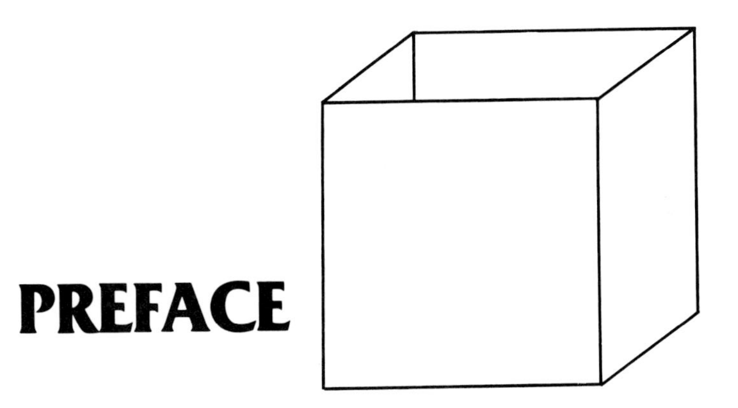

The unity here is functional form, which is intriguing, subtle, and beautiful. The essence is craftsmanship in a wide range of materials, skills, styles, and techniques.

We could not possibly have described in detail the legend and fact that are the tradition of each material, and we did not attempt it. Instead, each chapter describes, conceptually, the basic working processes associated with a particular medium. Step-by-step photographs of some of the world's finest contemporary craftsmen at work illustrate refinements in material and process.

Most important, however, is the rich variety of individual containers in each chapter. They represent varied solutions to similar problems, and show the application of materials and skills in both traditional and innovative ways. These forms not only survey the broad range of work being done today, but they suggest the medium's potential. We hope that they will inspire.

This book has been a joy to write, because of the people who contributed to it, the variety of materials included, and the pleasant (but often difficult) chore of selecting containers from work of outstanding range and quality. We hope you enjoy it too.

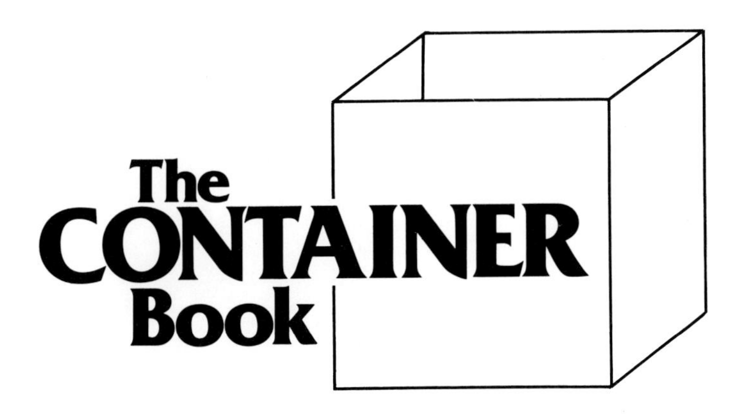
The
CONTAINER
Book

"Ekeko," the itinerant South American, would be lost without the bags, baskets, bowls, boxes, and leaves that contain his possessions and wares. He symbolizes "plenty."

Containers of many sizes and materials have been a part of every culture. Ancient Egyptian food offerings were contained in palm-leaf baskets, pottery jars, and wooden boxes. *Courtesy The Metropolitan Museum of Art*

CONTAINMENT POLICIES

1

People and containers are inseparable. There is a natural dimension to their relationship. Men are born and built of cells contained by specialized membranes. Complex conduits supply essential nutrients to discretely packaged organs. And an extraordinary wrapping—the skin—binds the parts securely.

People make containers too: as tools and objects they have been of considerable interest at every stage of man's social and aesthetic development. Containers are omnipresent in human civilization. Their roles as utilitarian forms, religious articles, articles of and for commerce, and works of art invariably reveal significant aspects of the societies that produced them.

From the first, people felt the need for containers. Once they learned that the elements could not be controlled, homes were necessary for protection. Very early, people felt the need for portable storage vessels. Nomads needed containers to carry their entire possessions. For foragers, baskets made food gathering more efficient. To hunters an arrow sheath freed arms to draw the bow. Animal hide sacks and bags protected precious shells and flints. Watertight skins released man from the limits of the life-giving stream, allowing him to explore.

As societies became more complicated, the range of tools and objects and clothing increased. With more stable settlements, the need for storage grew further; containers became larger and more permanent.

If the historical evidence is any indication, containers have also been something more than purely utilitarian forms. Even in societies where time for creative adornment is scarce, we find basic, essential, and handsome designs. Containers often exhibit man's attempt to produce functional and beautiful objects to enhance his environment, his life.

Early Containers: Natural Materials

Given the urge and the need to create containers, people have turned to the most prevalent and accessible materials available. Natural materials caught the container-maker's eye. The most significant of those substances found in nature were grasses, clay, wood, and metal.

Basketry

Basketry—the use of relatively large and rigid vegetable fibers to make containers and other forms—is an ancient craft. Plentiful materials—grasses, roots, shrubs, trees, leaves, reeds, and canes grow everywhere except the North and South poles—and the fact that tools are unnecessary to work these fibers combine to make basketry an ideal solution to many containment problems.

Despite the extensive use of baskets throughout the ages, archaeological evidence of baskets is scarce. Except under extraordinary conditions, such as extreme dryness or volcanic burial, the forms have not survived. Perhaps because baskets were so fundamental and commonplace, other records of them are also rare. Scholars, however, have been able to determine that the basic elements of basketry are the same today as they were five thousand years ago. Baskets are still coiled, built of wattlework and twined or woven threads, and plaited.

Contemporary Japanese baskets and boxes for food. *Courtesy Japan National Tourist Organization*

Above: "Ancestral Earn," handbag with screened enamel portraits of the artist's ancestors, by Vivian Kline. *Courtesy Vivian Kline*

Right, top: An antique Japanese basket.

Right: An American Indian basket (16" X 13") by the Tlingit tribe, from the nineteenth century. *Courtesy William Rockhill Nelson Gallery of Art, Atkins Museum of Fine Arts (Nelson Fund)*

Bottom, right: An ancient basket made by stringing together cloves, from Indonesia.

Below: "Victoria's Handbag" (45" X 36" X 36"), crocheted wool by Thomas Charles Siefke. *Courtesy Thomas Charles Siefke*

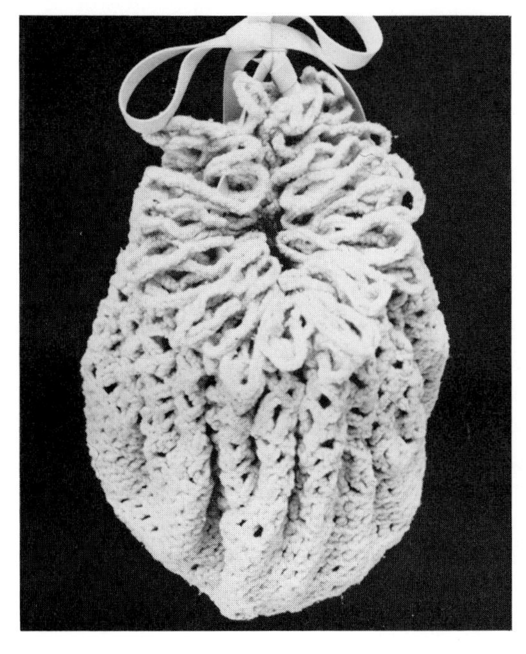

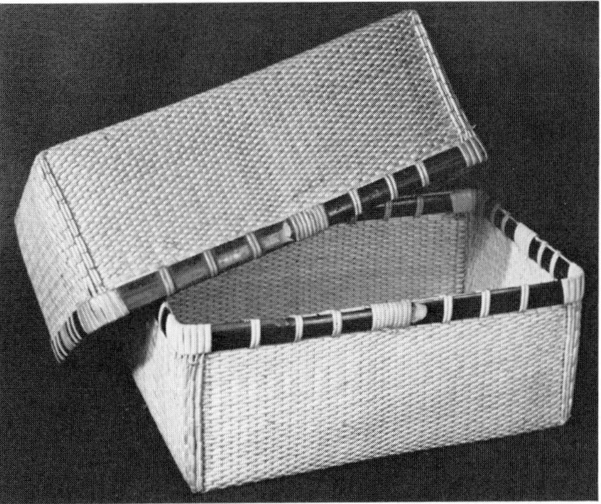

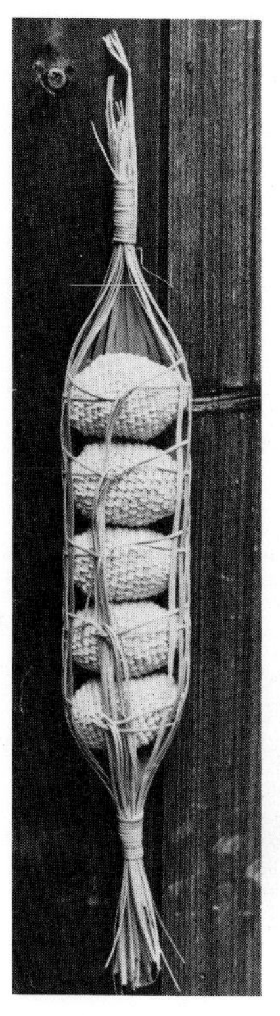
A contemporary adaptation of a Japanese egg basket—with coiled eggs, by Sally Edmondson.

Certain cultures are known for their emphasis on and skill in creating baskets. The American Indians probably developed the finest forms. Throughout the continents, north and south, individual tribes practiced basic coiling and plaiting techniques. But each group had its own signature. Tribes that inhabited what is now the western United States and Canada, the Chilkat, Tlingit, Apache, and Pueblo, did some of the most delicate work of all.

Of course, no group had a monopoly on technique and skill. In Oceania, plaited fibers composed ceremonial as well as utilitarian forms. And basketry was and is practiced in New Zealand, Polynesia, New Guinea, and Australia.

In Southeast Asia, bamboo is used in place of the reeds and grasses commonly employed in other regions. Southeast Asian basketry differs from that in other regions too in that its results are almost purely decorative and ceremonial rather than primarily functional. There, aesthetic concerns play an explicit and important role.

African people still rely upon baskets as utilitarian forms. But even though baskets are used primarily to carry large loads, store products and food, decorative elements are part of basket design.

Baskets are widely used for good reason. They are light, inexpensive, and often beautiful. The necessity for extensive handwork suggests that baskets could eventually pass out of the utilitarian domain, yet, the craft will probably continue, with baskets becoming luxuries, appreciated for their beauty.

There is no such thing as a typical basket, unless that is taken to mean a form large or small, made to carry things dry or wet, and loads heavy or light. Baskets are employed as plates and bowls; when woven very tight or coated with special materials they can be made watertight. They can even serve as cooking pots; one people boil water in them by dropping in hot stones. Openwork baskets serve as strainers for tea in Japan, and as grain sifters in other parts of the world. South American Indians construct tubelike baskets which, when stretched, compress manioc pulp placed within them.

As storage containers for objects small and large, baskets know no limit of application. Fine forms protect jewelry, others store clothing. Man-sized baskets have been used to store grain as well as to ship fragile goods on cargo ships. In harvesting, baskets are indispensable and especially well suited to the task since they add very little weight to the crop-gatherer. In hunting and fishing, basket traps are used to snare one's prey.

Not the least of the basket heritage, however, involves their use as ceremonial objects and vessels. In Bali, the vitality, variety, and complexity of basketry results from the ancient tradition of using baskets as offerings and containers for offerings in religious rituals and festivals.

A fifteenth-century leather Koran case embroidered in silver (lamellé). This bag probably belonged to Mohammed Abu Abdullah, the last Moorish king of Granada. *Courtesy The Metropolitan Museum of Art, Rogers Fund, 1904*

A French leather reliquary shoe from the fourteenth century. *Courtesy The Metropolitan Museum of Art, The Cloisters Collection, Purchase 1947*

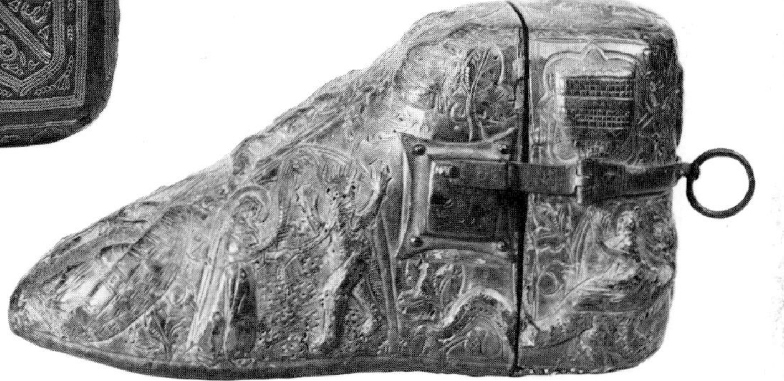

Egyptian marble and bronze toilet articles from the XII Dynasty. *Courtesy The Metropolitan Museum of Art*

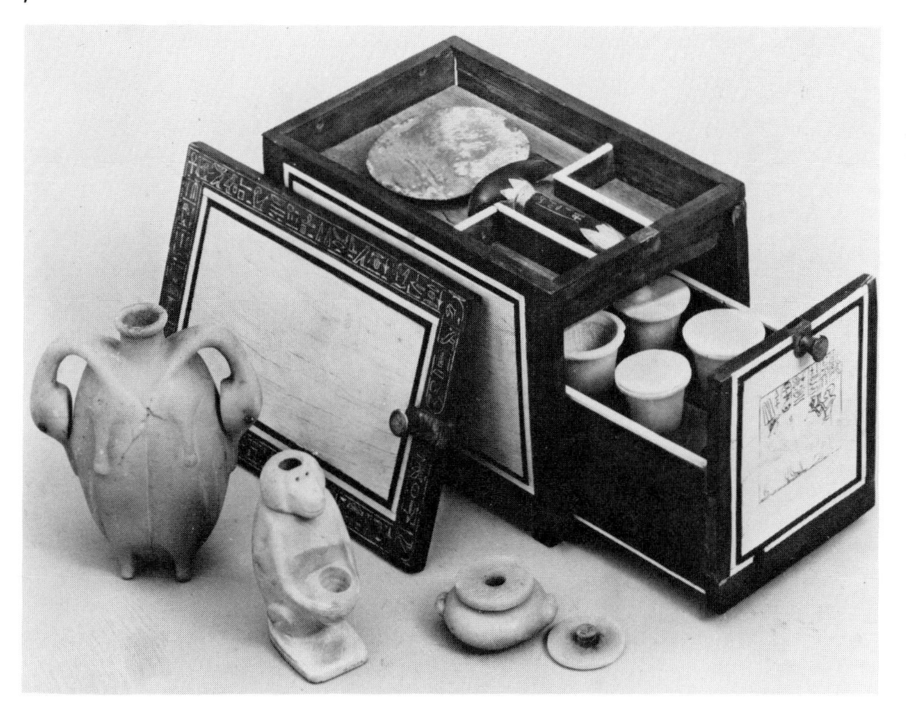

Clay

Crude, unglazed earthenware dates back to the Neolithic age, nine thousand years ago. Like baskets, ceremics rely upon natural—and naturally present—material. Clay can be dug from lakes, ponds, riverbanks, or deep deposits. Also, like basket fibers, clay requires few tools of any kind—and simple ones at best. There is archaeological proof that people discovered the basic working techniques quite early.

Ceramics can be built by fusing snakelike coils of clay; rolling out slabs, cutting pieces and pressing them together; molding clay over or into a form; "throwing" clay on a continuously rotating wheel, among other methods. By and large, these ancient techniques are the modern ones as well.

Once formed, clay may be decorated by pressing it, excising it, or coloring it with minerals, slips, and glazes. The hard, vitreous surface most common in the West is achieved by glazing the form and firing it in a kiln. In some areas of the world, ceramics are still fired on an open hearth, or in a bonfire, while more sophisticated potteries employ walk-in kilns with carefully controlled heat and ventilation.

The real story of ceramics, however, is individual perception. Like every decorative art, and like every creative enterprise, ceramics is a worldwide phenomenon as varied as the people who create the forms. From the unglazed Neolithic pot stemmed much more complicated designs.

The next step occurred in Egypt and Greece where pots with ornamental slips and molded embellishments appeared. Later, in Greece, the high art developed. By the later Bronze Age, some 3,500 years ago, abstract motifs, varied color schemes, and sophisticated shapes had developed. In the Archaic period, the renowned black-figure and red-figure ceramics were thrown.

At the same time, the Far East had developed its own clays, glazes, traditions, and processes. In time, each dynasty and each Oriental culture found a ceramic signature of its own.

But tradition and craftsmanship were never the exclusive province of the Orient or ancient Greece. The Etruscans, Romans, and Persians maintained distinctive forms and decorative motifs of their own. And in the Americas, magnificent pre-Columbian forms are still being unearthed at grave sites.

In the West, ceramics have had an equally vigorous, if comparatively shorter, history. Sèvres porcelain, Meissen stoneware, Delft tiles, agateware, jasperware, majolica, and bone china are but a few Occidental contributions to the craft.

A stoneware mug facing left, by Dave Boronda. *Courtesy Dave Boronda*

"Gothic Lady" (8" X 9" X 6"), a porcelain dream box by Paula Winokur. *Courtesy Paula Winokur*

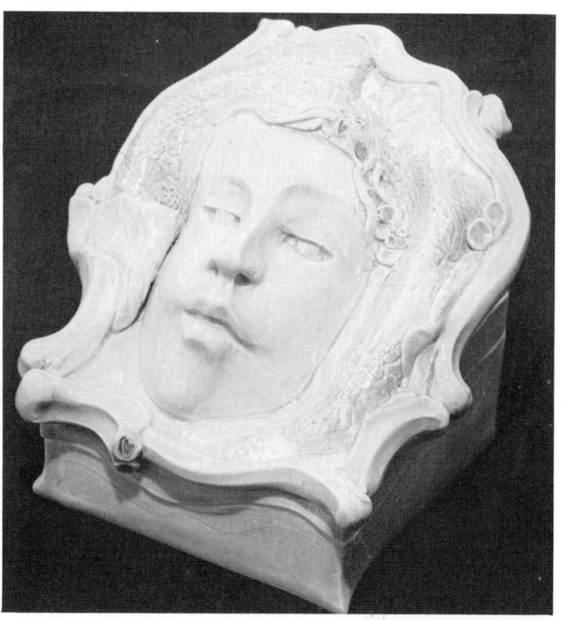

A rainbow kettle by Judith Rothbart. *Courtesy Judith Rothbart*

"The Athlete" (13" X 10" X 20"), ceramic cookie jar with underglazes by Joyce Moty. *Courtesy Joyce Moty*

Opposite: Small stoneware mugs with riveted lips and luster glazes.

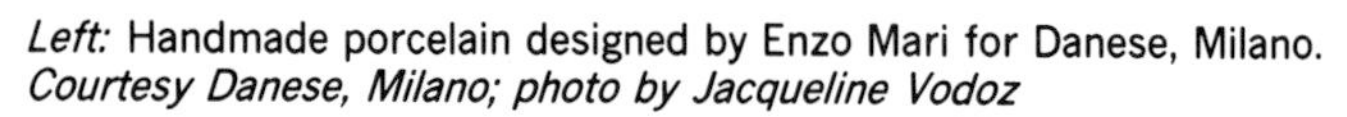

Left: Handmade porcelain designed by Enzo Mari for Danese, Milano. *Courtesy Danese, Milano; photo by Jacqueline Vodoz*

Below: A Byzantine carved ivory casket (3⅞" X 13⅝" X 5¾") from the eleventh century. The bronze fittings are probably of a later date. *Courtesy William Rockhill Nelson Gallery of Art, Atkins Museum of Fine Arts (Nelson Fund)*

Bottom, right: "Entomology Neckpiece #1," by Jan Brooks Loyd, of metal and glass, containing insects. *Courtesy Jan Brooks Loyd*

Bottom, left: A glass vase (17" X 10" X 3"), by Kent F. Ipsen. *Courtesy Kent F. Ipsen*

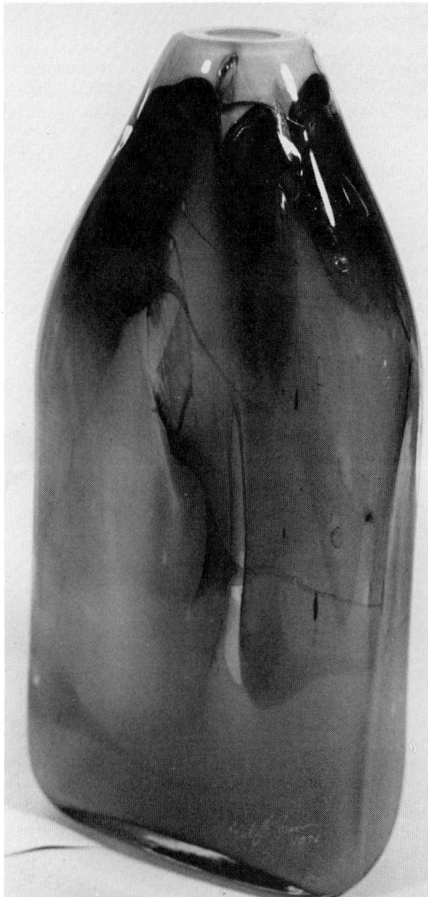

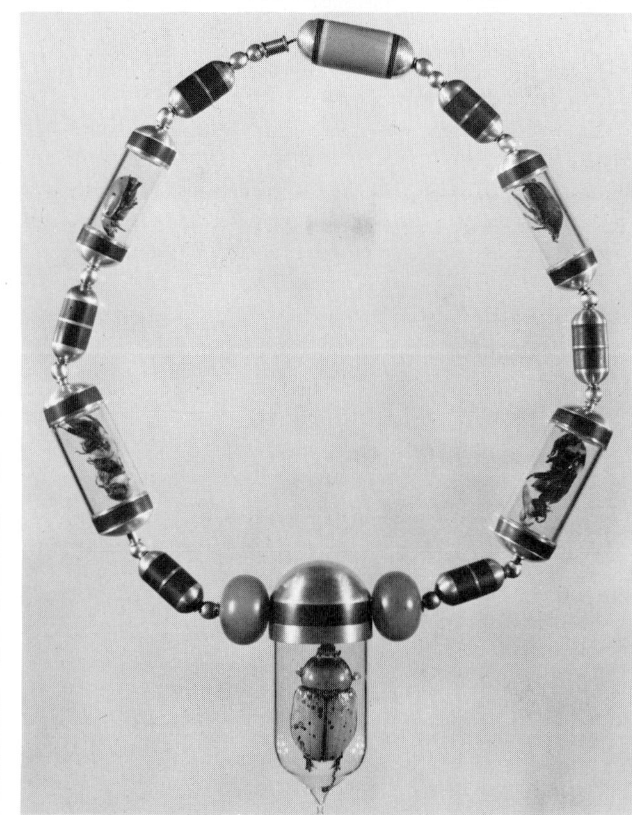

Wood

Wood—like clay—probably emerged as a structural, functional material during the Neolithic age. For the first time, with ground and polished stone tools—ax, adze, and chisel—people could control the trees in their environment. They became capable of felling woods to clear fields for planting. Furniture, bowls, boxes, coffins, structural elements, and even other tools were fashioned with considerable precision.

With woods of different textures, colors, and consistencies, the container-maker had a vast resource. The use of wood in containers and storage forms has been extensive—wherever material less fragile (than clay) and more permanent (than vegetable fibers) was needed.

In ancient Egypt, finely painted wooden boxes served as both furniture and storage units. Thin sheets—veneers—of wood were glued together to build coffin cases. Wood boxes were an element of Mesopotamian furniture and storage design too; there, craftsmen carved intricate designs in the panels.

The Greeks also built wood boxes. Romans stored their clothes and money in them, and built elaborate treasure chests of wood clad with bronze and iron. Small items and jewelry were stored in more portable round and square boxes.

Particularly during the Middle Ages, wood boxes were a necessity. Furniture was quite scarce, and what little there was had to serve many functions. The medieval chest, constructed of wood planks nailed or pegged together and reinforced with iron, was the basic unit of furniture, storage, and transportation of possessions. It proved so versatile as cupboard, trunk, seat, and desk that the box chair was developed. Until the late fifteenth century, this large crude chest remained the single most common and useful item for medieval man with possessions. As communities stabilized, the need for fixed, nonportable storage systems increased and the cupboard appeared.

During the Renaissance, cabinetry skills developed into high art. The styles of that age were developed, embellished, and propagated by skilled woodworkers through furniture design. Ornately carved chairs, chests, couches, and beds emerged from workshops.

In the seventeenth century the bulky chests of an earlier period were succeeded by chests of drawers—bureaus. More people possessed more items that required storage places. Again, master woodworkers applied their skills to the problems of storage and beauty and, again, new forms resulted.

Antique Japanese silver saki cups encased in a wood box with sliding top.

Contemporary Japanese packaging of handmade, hand-printed paper, and wood.

Traditional nested Russian container for a container in a container toy.

Japanese bronze and silver "Inros," portable medicine cases, from the eighteenth and nineteenth centuries. *Courtesy The Metropolitan Museum of Art, Bequest of Edward C. Moore, 1891, The Moore Collection*

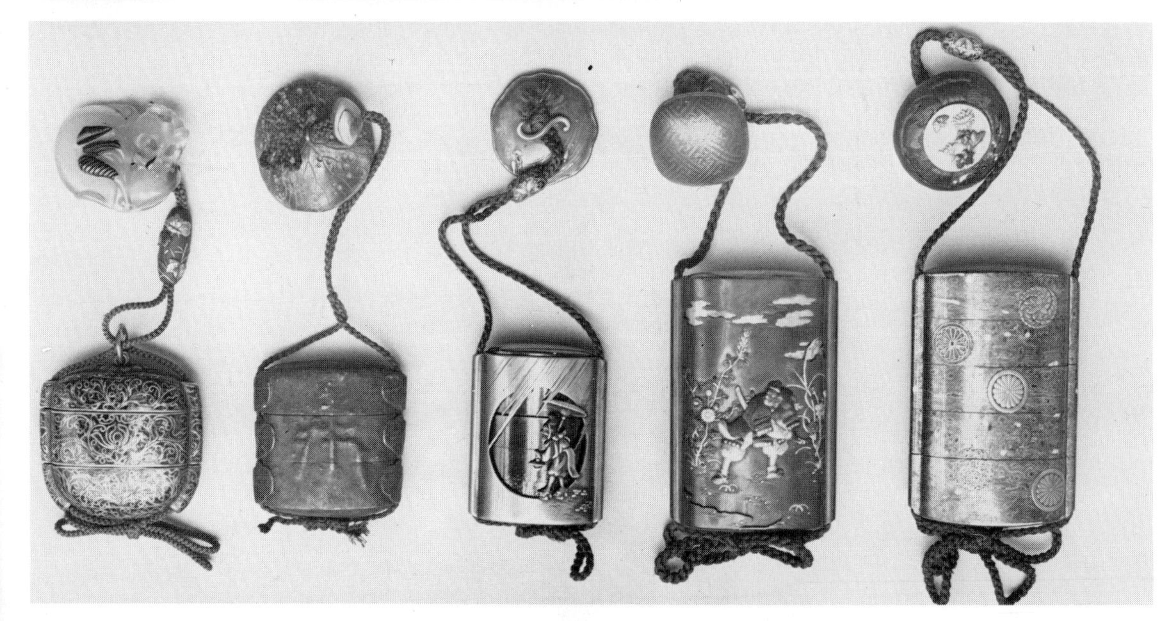

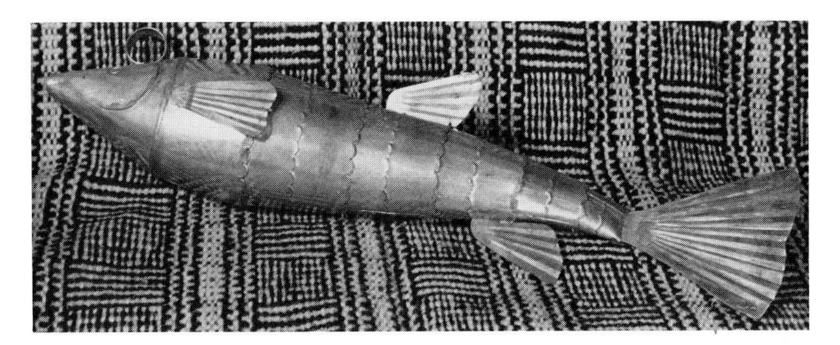

Ceremonial liquor container with articulated sections in the shape of a fish, from Peru.

"Treasure Chest" (7" high), in calfskin, walnut, and cast and formed sterling, by Earl Krentzin. *Courtesy Earl Krentzin*

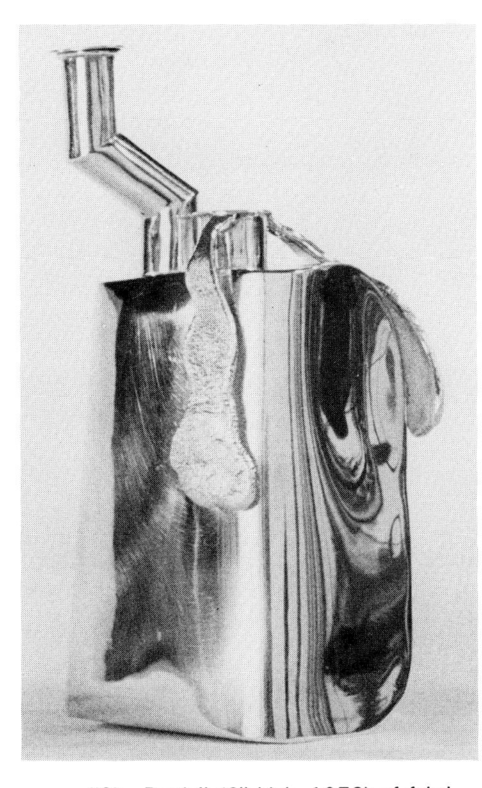

"City Bottle" (6" high, 1973) of fabricated sterling silver by Margaret Laws. *Courtesy Margaret Laws; photo by J. Bryant.*

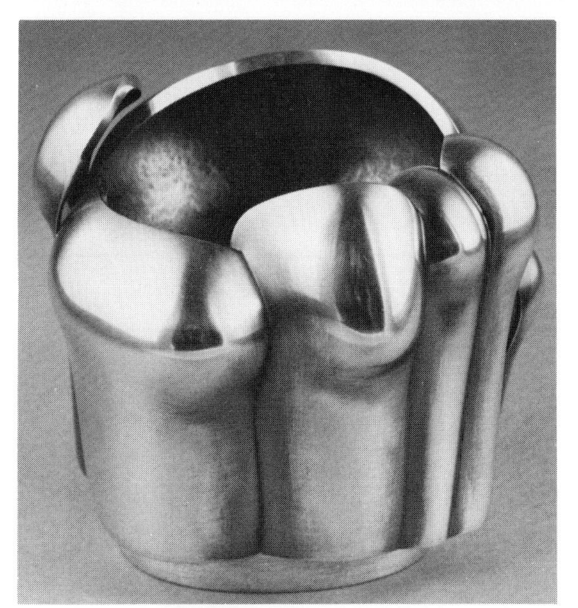

Overlapping forms container (3¼" high) of formed and constructed 18-gauge sterling with an oxidized finish, by Helen Shirk. *Courtesy Helen Shirk*

Left: "Commemorative Pendant to the Farmer's Daughter for All the Dirty Jokes about Her" (6" x 1"), in silver, plastic, and dirt, with flag, by Jim Cotter. *Photo by Tom Lamb*

Right: "Shoulder Bag" (8" X 24" X 1½") in silver plate, calf, and psilomelane stones, by Marcia Lewis. *Collection of the Museum of Contemporary Crafts; photo by Dennis J. Dooley*

Bottom, left: Daniella Kerner's "Cocktail Canteen" (10" X 5" X 4½"), of sterling silver, Delrin, and vinyl tubing, was electroformed, lathe-turned, and fabricated. *Courtesy Daniella Kerner*

Center, right: Shell box of shell and fabricated silver, by Marguerite Stix. *Photographed at the Kruger Van Eerde Gallery*

Bottom, right: Cast aluminum roaster (12" X 8½" X 4½") by Don Drumm. *Courtesy Don Drumm*

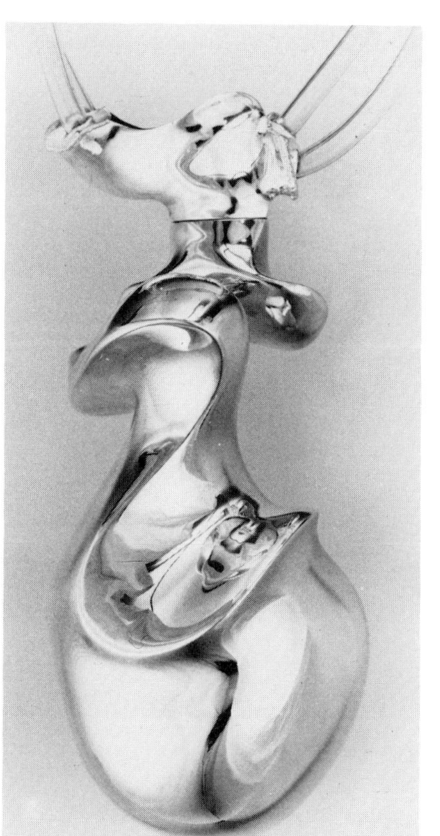

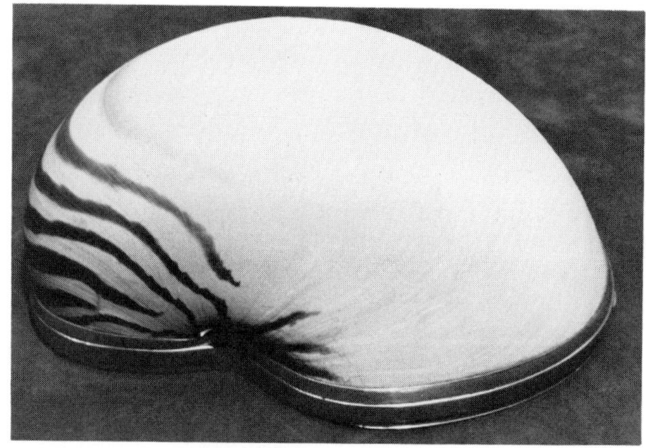

Metal

While vegetable fibers, clay, and wood were tapped by container-makers early in human prehistory, the extraction and working of metals represented a later stage of development. But again, once the elemental methods of fabrication were discovered, everything else—virtually to this day—was largely a refinement of technique.

Lost-wax casting and metalworking with hammer and punch were two such processes discovered by container-makers. These techniques reached a high level of refinement in various parts of the world.

The first metal commonly worked by metal craftsmen was copper. Vessels of raised copper were executed in Sumer during the 4th millenium B.C. Egyptian copper basins and bowls followed slightly later. Since that time, craftsmen have often turned to copper for a malleable, relatively inexpensive material, and even during the Middle Ages, metalworkers engraved copper religious objects.

"Form #108" (8" X 8" X 13"), of copper, electroformed, milled, and carved cast acrylic, by Daniella Kerner. *Courtesy Daniella Kerner*

But for certain purposes, copper was too malleable. Craftsmen soon discovered that the harder alloys of tin and copper (bronze) and zinc and copper (brass) were much better suited to casting.

While copper, bronze, and brass were used extensively, the very finest work always appears to have been done in precious metals. Of course, the definition of precious has changed with time—the Roman nobility sterilized themselves by drinking and eating from "precious" lead vessels—but it seems that during most of human civilization, gold and silver have been loved and lusted after most often.

Over four thousand years ago the Trojans executed vases in silver and gold. Grecian bowls, decanters, cups, goblets, and ritual vessels date from the same age. Animal and architectural motifs decorate Greek forms, while Roman embossers favored mystic, mythological, and historical scenes.

The skills that were not lost during the Middle Ages were used by Christian craftsmen to create religious forms. By decree, sacred vessels could be executed only in precious metals, although several formed in baser elements have survived. Censers, chalices, bowls, basins, and plates were created during that period. In fact, with the church as a major social force, much of the work was done in monasteries, where skills were preserved during the Dark Ages.

But as that period shaded into the Renaissance, metalworkers throughout Europe began producing eleborate containers for the burgeoning royal courts and the slowly emerging middle class. In time, English and American colonial silversmiths adapted those skills to new traditions in design and

Lisa Drumm's doll unzips to contain another. *Courtesy Lisa Drumm*

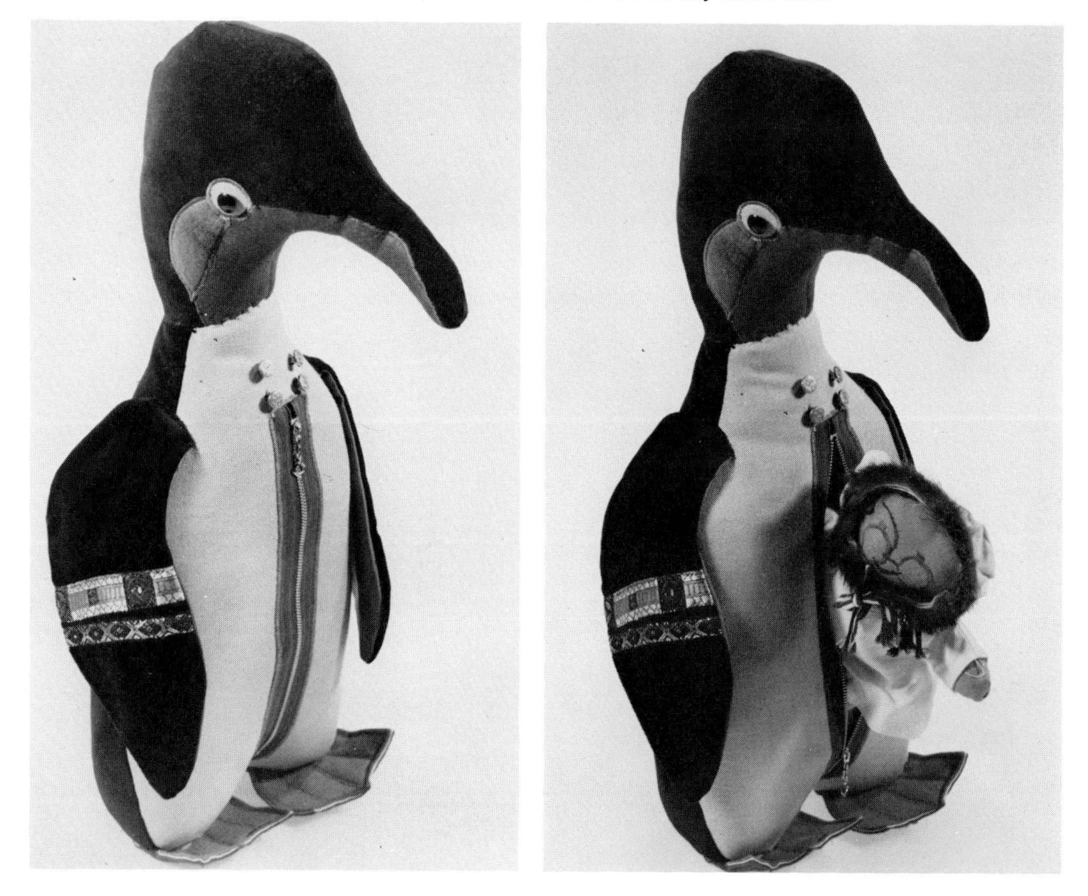

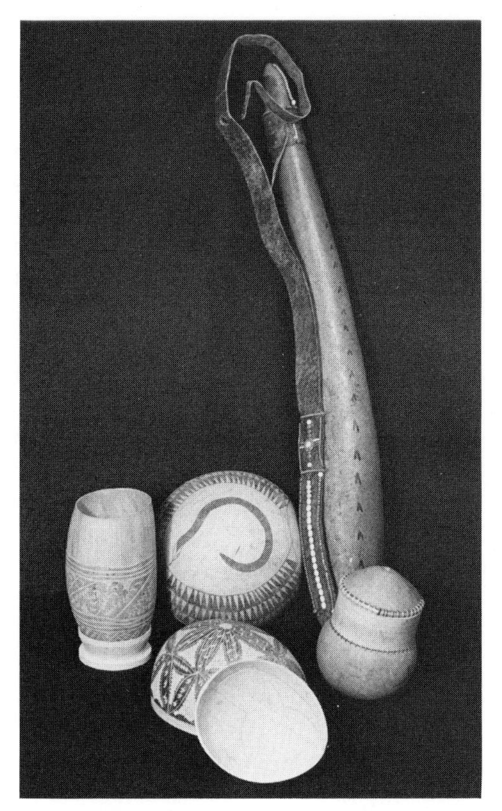

Gourd and calabash containers from Africa and South America.

function. The colonials, poorer than their old-country cousins, did much of their innovation in pewter.

The story of fiber, ceramic, wood, and metal containers carries much more with it than mere description of the forms and techniques can convey. Implicit in each container is a social order, a minute reflection of a civilization. In the periods discussed above, for example, all containers were created by individuals. Early societies might not have been so specialized that they needed one person's efforts devoted to a particular craft, or the craft, like basketry, might have been something that everyone could do or had to do in order to transport goods in their nomadic wanderings. But, for the most part, the story of civilization is the story of specialization. In most societies, individuals develop skills in certain crafts and become so expert that members patronize those craftsmen.

Also implicit in the social concept of craft is the notion of prestige; it enters from two sources. From the craftsman's point of view, the more precious, expensive, or rare the skill—the more difficult and demanding the work—the higher society values it, the higher that craftsman stands in that society. On this scale, the basketmaker and the potter occupied the lowest rung of the social ladder. Pottery and baskets broke and had to be replaced. Performing essentially (or apparently) simple tasks over and over again, their work was regarded as more of a service than a skill. It seems that, traditionally, metalsmiths, and especially gold- and silversmiths, were the most highly regarded (and highly paid) of craftsmen. Skilled woodworkers were valued because of their contribution to the artifacts of the church and the homes of the wealthy. These materials lasted longer and commanded a higher exchange rate.

The other point of view is that of the consumer of a craftsman's work. Different societies, of course, define success and status in very different ways. In some Western cultures, people define themselves by wearing other people's initials. During the Middle Ages, men and women used elaborate purses called *aumônières* to define their station. The most royal had these purses ceremoniously carried by servants. In any event, what society valued little, it had little interest in preserving. Ceramics lent themselves to mass production. Another workaday craft was basketmaking. Both were swamped by the wave of industrialization and that new form of container—the package.

The Package

The prehistory of packaging, of course, is the story of natural materials worked by skilled craftsmen. In man's search for order and protection of his possessions and person, he developed an extraordinary range of vessels. The transition from container to package occurred gradually, over several hundred years.

Almost by definition, there could be no packaging before there were trading communities. People always traded, but the type of industry that required packaging was one that depended less upon individual, face-to-face contact to sell a product and more upon the building up of a name or symbol for a product, without need of direct human contact. One of the earliest such trades was German papermaking. During the late sixteenth and early seventeenth centuries, the papermakers began to market their more or less standardized products more widely than the towns and villages in which the papers were produced. Ornate, descriptive labels attracted customers and described the contents. So, too, did the labels used on the patent medicines advertised in London newspapers during the late 1600s. But even during the seventeenth century packaging was not the universal mode, as witnessed by a newspaper advertisement that suggested that patrons bring their own boxes to contain tea they purchased.

Except for some companies—Yardley, Twinings, several beer manufacturers (and bottlers of beer antidotes)—few in the eighteenth century made use of packagings or brand identification. Yet, there were good reasons for not creating distinctive and descriptive packagings: literacy was rare and most firms were still predominantly local. In addition, technology had not yet developed to the point where large quantities of packing materials could be produced effectively.

With the turn of the nineteenth century, several coincident events foreshadowed and encouraged further changes in modes and methods of containment. The first was the invention of a papermaking machine; the second, the invention of lithography; and the third, the initiation of American packaging.

The first two really represented the economic aspect of packaging. Robert's papermaking machine made it possible to produce papers of a uniform quality quickly and inexpensively. Although papermaking had been rediscovered in the West several hundred years earlier, it was a laborious process until the nineteenth century. Senefelder's lithography process revolutionized many modes of communication. Lithography offered a very quick and inexpensive printing process, perfectly suited to labels, cartons, and the like, once paper was available in large quantities.

But the third aspect of this trio of innovations is more difficult to explain. The development of the American packaging industry—soon to include a wide variety of machines to make packages, measure and weigh the contents, fill and seal them, and affix the label—resulted from something much more basic. The need for uniform packaging, in quantities large enough to require complex and fast machinery, arose from the nationalization of marketing and manufacturers. That is not to say that the government took control of the companies, but, in a real sense, companies began to take control of the country. While manufacturers had previously produced for local markets—which, because of the structure of American society, were small by definition—development of large urban centers produced an incentive to expand and specialize. If a town of one thousand became a city of ten thousand, local producers obviously had an incentive to produce enough to supply the entire populace. While in smaller towns their products were readily known by word of mouth, such personal communication and knowledge became much more difficult in larger centers. To reach more people, and perhaps even old customers, it became necessary to affirmatively market and identify products. Uniform packaging—bespeaking uniformity of quality—provided one means of communication. (Extensive advertising provided another.) And once manufacturers and traders discovered they could reach an entire city, or state, it became apparent that they could, with only slight additional effort, reach an entire country. Hence, a revolution in marketing, and the appearance of the brand and the brand name. Before the twentieth century, however, packaging was still not what we know today; by and large, packaging pioneers adapted ancient crafts to the manufacturing process.

The ancient potter's art was transformed into a modern packaging mechanism with the introduction of the transfer-printing process. After all, hand-lettered ceramics had little place in the factories where packaging materials were mass-produced. Transfer printing enabled the product manufacturer to impart scenes as well as product names and descriptions onto their vehicles. The process was a simple one, too. The design is first printed on special paper from an engraved plate, or lithography stone, and the paper is applied to the pot in its unglazed state. When the ink dries, the paper can be washed off, and during the firing process the oils are burned out of the ink leaving a metallic oxide on the surface of the ceramic. At first, transfer printing was done only in a single color, but multicolor printing followed shortly. At least one English company—Crosse & Blackwell—received applause for its fine Wedgwood pickle pots. Of course, still other companies employed printed paper labels that were affixed with glue.

Analogous to ceramic packages were those of glass. Quite often they were molded with lettering and decoration, much as the Rose lime juice bottles are molded this very day. Paper labels were readily affixed to the glass as well. Glass first emerged as a container for beer, cider, and wine. With the development of automatic bottle makers, their use became even more widespread.

Paper—the universal packaging material today—began its service as a wrapping and as a printed label. Soon, however, many products, from cough drops to tobacco, cigars, and Worchestershire sauce bottles, were enclosed in it. One Tom Smith introduced the world to mammoth consumption of paper, cellophane, and tinfoil with his invention of individually wrapped candy.

In Bristol, England, the paper bag, printed and plain, emerged. It permitted easy wrapping and advertising, features that account for its popularity and utility to this day.

The heavier paper products—cardboard, corrugated board—were, of course, part of that age of expansion in marketing too. With Dennison's invention of machinery for creasing and cutting boxboard mechanically, the modern era had been reached. What paperboard boxes replaced were, very often, wooden slat crates and round or oval containers constructed from compressed wood shavings.

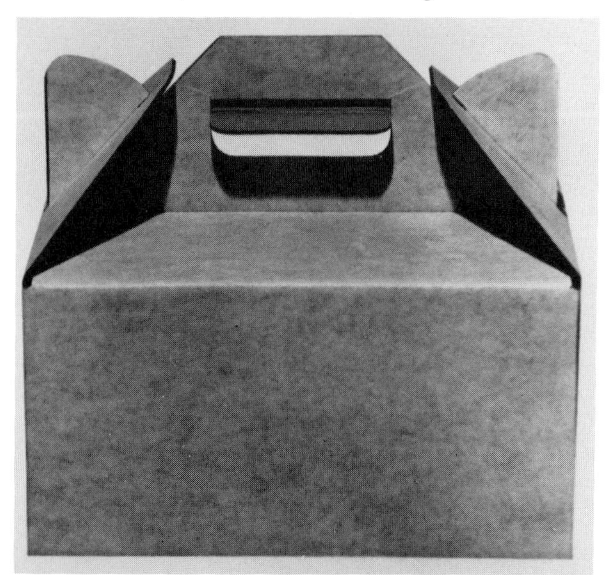

Pet container of solid fiberboard, manufactured by the Container Corporation of America. *Courtesy Container Corporation of America*

The next step, accompanying the large use of boxes for products and shipment, was the folding box. Rather than take up the space required by the quantities of boxes necessary to ship large quantities of a product, folding boxes economized; they could be erected as necessary. By 1897 over eight hundred patents for folding boxes had been issued.

Probably because metalworking required high degrees of skill for so many centuries—dating from boxes made by Sumerian and Egyptian artisans—metal packaging did not emerge until the mid-eighteenth century. Its first application was a tin used as a tobacco canister. The biscuit makers probably noted that, if tin was good enough for fragile tobacco, it could keep crackers crisp too. They followed suit. Tin provided a fine, solid, airtight vehicle for citizens of the empire living at its far reaches.

Early metal boxes, like the first papers ones, were decorated with paper labels. Some packagers soldered thin metal labels onto the tin. Clearly, if it were possible, the best solution would be to apply the ink directly to the metal. Transfer printing was first used for that purpose until offset lithography took its place. Tin printing soon became an industrial process in itself.

Food cans represent another aspect of metal containers in packaging. Nicolas Appert first developed them in the early 1800s to feed Napoleon's armies. After experimenting with wide-necked glass jars sealed with corks, he developed the cylindrical tin with a soldered top. Appert still apparently preferred glass for most applications, regarding tin as a substitute only for extreme conditions—like the battlefront or ship voyages. But by 1840 tin cans were widely used in Britain and the United States. From a production rate of one a minute by hand during the 1870s, factories were sealing 2,500 an hour by the turn of the century.

The Plastic Revolution

However revolutionary national distributions were on the artisans' container industry, the impact of plastics was many times greater. In less than forty years polymethylmethacrylate (acrylic), epoxies, polyethylene, polyurethane, and polycarbonate have changed the face of the world. No matter what material was used before, it seems the plastics industry has had an alternative solution. Because the price is often less—or the benefits greater—plastic containers tapped a large market.

Of course, the disasters along the way are familiar to many. Early plastic products were often poorly designed. It took many years for plastics to live down the derogatory tone that became part of its name. But, as the industry and design community have matured, plastics have found successful, perhaps indispensable, applications in nearly every facet of life. One reason for this popularity is their versatility. Plastics are available in sheets, rolls, liquids, and pastes, and can be formed into virtually any shape imaginable—

by cutting, casting, foaming, molding, stamping, heating, and a myriad of other techniques. Packagers have found plastics to be an extraordinary resource.

The advent of industrialized packaging and widely transported products has not, however, been all for the best. In many ways, the use of disposable packagings—which most are—is an ecological disaster. Forests are cut to provide paper pulp and, although they are truly farmed to produce trees, one cannot help but wonder whether something has not been lost. The use of plastics in packaging poses problems of disposability: while all paper is biodegradable, most plastics are not. Perhaps the attitude that caused tea merchants to advertise that customers had better bring containers for their tea represents a more economical and socially responsive view. Container, rather than package, may be the more responsible course.

Containers and Packages

Implicit in this discussion is the notion that a package is somehow different and distinct from a container. That may be semantics, but there is something more permanent about a container. The distinction is meant to point out that packaging, with the possible exception of the Japanese art of wrapping and package design, is an industrial process. Packagers have little interest in the finer elements of design, construction, and craftsmanship, unless these relate to the consumer-mind image of product. The reason for drawing this distinction is the renaissance of interest in artisan skills and the rekindling of interest in finely designed beautiful handmade forms.

That is the reason for this book, to show what people can make, what they have made, and how they work. Artists and artisans throughout the world have turned to containers at one point or another because the vehicle is so much a part of the human story. No matter how nonmaterialistic, no matter how ascetic, people need containers, and for the most part they want them to be handsome as well as functional.

DESIGN FACTORS IN CONTAINER-MAKING

Influences of Form

Social conditions are not the only factors influencing the development of container forms. Design has also been influenced by nature's own containers —materials available in the immediate environment, and by the intended function of the container.

The earliest models for containers must have come from nature. Birds' nests are baskets for eggs. Wasps' nests are paperlike coverings to house and store. Holes in trees shelter small animals; shells of turtles and fish such as scallops are portable, protective, water impervious containers for soft fleshy life. Forms that grow on trees and on the ground—gourds and calabashes—are another natural container source. When dried, the resulting hard, rigid skin can be cut and decorated to create highly valued waterproof receptacles. In some West African marketplaces, skilled craftsmen mend precious calabashes on the spot.

The wonder is why people who had to work from sunup to sundown just to survive expended additional effort to color and decorate their containers. Apart from obvious functional requirements such as lids, knobs, handles, and closings, which do embellish even purely functional forms, decorated containers were symbols of pride in skill, objects of ritual and protection from unknown forces. Beads, feathers, embroidery, gems, and patterns imparted status to otherwise banal objects, and created a place in the hierarchy of what is loved and treasured in society.

It is not enough, then, to say that a container is defined by purpose and material alone. The form also must be beautiful in order to survive the test of time. And what makes a form beautiful is the most difficult question of all, because the answer changes from time to time, as do styles and fashions. The containers that have survived use and the elements—those that are still treasured—have some common aesthetic denominators.

Nature reinforces a rather thin fibrous covering with corrugations caused by stringlike growths in the peanut's shell. Each nut, contained in its own cavity, is also wrapped with an additional paperlike cover.

Barnacles are container homes for sea life. Each cavity builds on to the next for strength and protection.

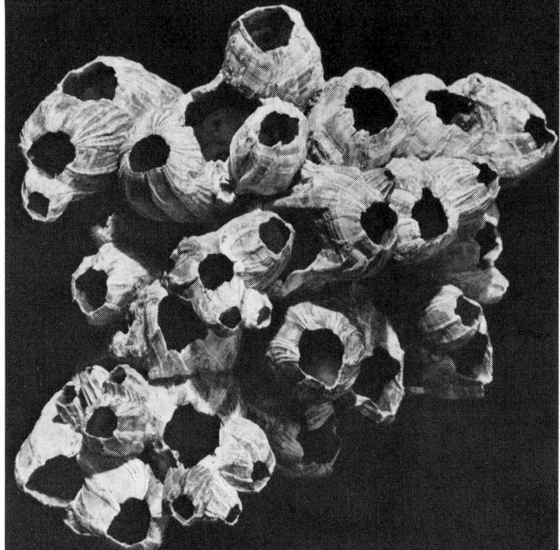

Another of nature's marvelous containers surrounds and protects the food of life, acting as a waterproof container that is quite strong yet can be broken easily when necessary.

Section of a pearly or chambered nautilus *(Nautilus pomphius),* a hard, protective waterproof outer covering or exoskeleton.

The scallop shell, corrugated to increase its strength, utilized as a compact. By Marguerite Stix. *Courtesy Kruger Van Eerde Gallery*

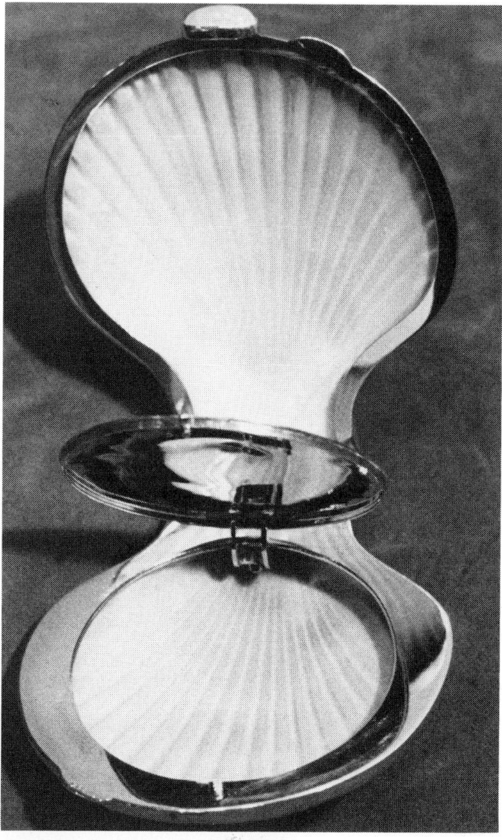

Coconut shells converted for use. On the left, nested with raffia, by the Toradja of Sulawesi (Celebes), Indonesia. On the right, one from the Kalinga people of West Africa, decorated by incised lines in geometric patterns.

Common Factors of Successful Form

Control of Material and Process

Unlike manufactured products, most craft forms are nurtured by craftsmen from the acquisition of materials to the final sale. Sometimes the impetus to build a particular form is motivated by the external demands and needs of the marketplace—what people will buy. At other times a design idea begs for expression. Sometimes a design is commissioned. But in all cases, there is latitude for choice on the part of the artisan. From inception, the craftsman/designer has control over the selection of materials and the creation of the form.

The craftsperson chooses materials and working processes. The most successful craftsmen are those who evaluate not only a material's potential, but whether its intrinsic and extrinsic qualities will be respected in a particular design. In some cases, appearance is the only consideration. One example is the use of polyurethane, a plastic with a wide range of potential application as a structural material. Although it can be and is used in many ways, its chameleonlike ability to imitate other substances appears to be its most popular mode. Significant percentages of contemporary "wood" furniture are not wood at all, but cast plastic exactly duplicating wood to the very qualities of texture, grain, and color. Those products have no integrity: materials should be used for what they are. Polyurethane *can* replace wood, but why should it be used that way when it has qualities and potential all its own?

Craftspeople who follow the production of a form through all its steps can control all factors. If a craftsperson took a beautifully grained piece of walnut and painted it with a solid color glossy enamel, it would be ludicrous, a blatant loss of control over process. When a thrown piece of ceramic is allowed to dry unevenly, resulting in warping and cracking, it is bad technique and another example of loss of control—more obvious than the polyurethane furniture which would require a close examination to distinguish from wood.

Form and Function

Control over material and process are attributes of the form/function design component. The material to be used, and the way it will be processed, depend in large part upon the function of the piece. For example, a detailed ceramic receptacle would be better executed in finely textured porcelain than in a heavily grogged (coarser) stoneware.

In other instances, a container form is just a vehicle for a sculptural idea. Function has little or no meaning, and form alone becomes the dominant factor. This kind of fantasy—to which function is irrelevant—is just what craft forms need. It is a new vitality. Heritage from the past is often limiting. Although much can be garnered from classic forms born of a tradition, too much dependence upon rigid, standardized design concepts can deaden and stultify innovation. It should be kept in mind that although form and function

A wire salad basket allows water in and out while gently enclosing fragile "greens."

Flexible baskets designed to store and protect eggs while allowing air to circulate through the container. The one on the left is Amerindian from Surinam, South America; on the right is a basket from Quibdó, Colombia, South America.

Far left: The outer shape of this form became a function of its "container-ism." "Sewing Scissors, Needle Case and Thimble" by Michael B. Riegel. *Courtesy Michael B. Riegel*

Left: The scissors as they relate to the five-inch-high case.

Below: Looking into the case. The materials are mild and carbon steel and brass. The forms were made by forging, raising, drawing with a flexible shaft tool, and fabrication. *Courtesy Michael B. Riegel*

"Knitting Needle Bag" (6" X 11") by Norma Minkowitz is knitted on circular needles. The bag, supported by a metal armature, was specifically designed to contain knitting needles. *Courtesy Norma Minkowitz*

Some fantasy along with function in Joy Nagy's "Satin Sandwich" (6" X 5½" X 2" deep), which is also a purse. *Courtesy American Crafts Council; photo by Bob Hanson*

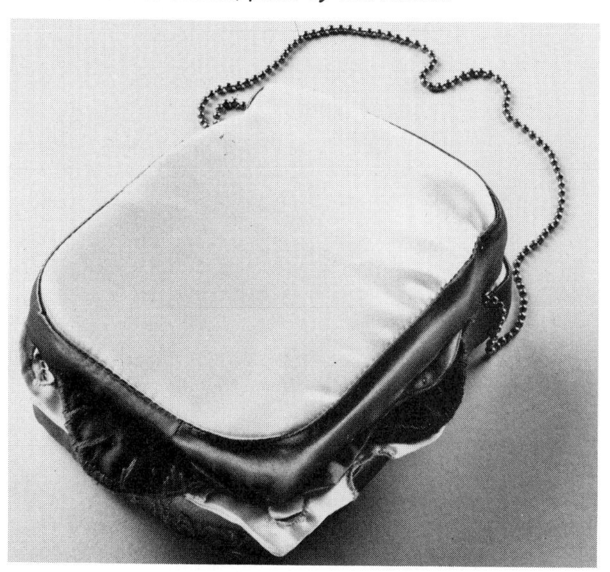

are vital considerations, vessels of the past do not necessarily belong in today's environment.

In the twentieth century, the first industrial designers were architects and stage designers. They brought another dimension in design to mass craft. As designers for first-class mass-produced manufactured products, they developed pacesetting forms, designs that led the way in exploring the potential of the media in which they worked. Because of inputs from other disciplines, the old saying "forms follow function" has been expanded. We have learned from new technologies and new materials. Although sometimes high-technology manufactured forms are insidious influences, they are too persistent to ignore. Moreover, there is still much to learn. On-site foaming of polyurethane, with its free, bulbous, overlapping planes, could have influenced the building of large ceramic forms. Phototechnology has certainly provided a new dimension to surface treatments and subject matter in the design of many forms. Of course, the form/function relationship has also been modified by the needs of new living styles, and the competitive economic relationships between products in a "free" society.

In the United States, plentiful raw materials have fostered waste and forced obsolescence. Today, this attitude is changing because artists and craftsmen do not want to think of their personal creations as disposable or short-lived, and because of the growing awareness that resources can and are being depleted. This new consciousness has resulted in increased consumer respect for handmade forms. As more people adopt this long-range view, the handmade form will become even more popular. Well-made, well-designed forms are more economical even if the purchase price is higher; they last and do not become outmoded with the stylistic changes in mass-produced forms.

Balance between traditional aesthetics and machine-age designs can be maintained if consumer judgments are educated to product quality. Poor design, poor workmanship, and improper use of materials are a distinctive phenomenon; they eventually lead to demoralization and the sense that nothing is good or right. In the long run, this is a symptom of a culture's disintegration. The goal is to preserve and treasure past achievements, and maintain our standards without stultifying innovation. The novelty item and the product designed for its own sake upset this balance. Thus, container designers must consider form/function aspects of past craft history, contemporary social and economic requirements, and new technologies and materials.

Technique and Skill

The degree of skill in any medium increases the expressive potential of the craftsperson. The craftsperson who has successfully mastered (in depth) the mechanics of working in one medium can also transfer this methodology of design to new and different materials. The same insights and processes of organizing and building ideas into real forms take place. Of course, each material involves a new vocabulary of process, but adjustments to new areas can be made with great success once one has been mastered. Transferral of skills and concepts becomes easier and easier with the accumulation of experience—both in new media and in making the transferral.

Plastics as plastics. Clear and opaque acrylic vases by Neal Small. *Courtesy Neal Small*

It must be emphasized, however, that what is valid in one medium is not necessarily adaptable to another. Each material has limits on how it may be worked. In different materials the form may change although the function remains the same. A typical error of judgment was made by manufacturers of early plastic products when they imposed designs for wood on plastics. They compounded the error by mismatching plastics to particular applications—resulting in fragile products. Plastics still suffer a bad name as a result, even though they have become a successful and ubiquitous element of our environment.

Each tool and machine, each material and process, has its intrinsic qualities and special concepts related to its use. These qualities should be respected lest the product's integrity bear the brunt of misuse. Respect for material and process is vital for successful designing of containers in any medium.

Design Planning

Design planning is basically a self-questioning and structuring process. How are you going to design the piece? Why use a particular process? What should the size and shape be? How will the piece be finished? And so on.

Designing is the whole of conceiving and making a container. It may require selecting, rejecting, experimenting, and testing until the most satisfying solutions emerge. To begin, define the problem—what are you trying to accomplish? Find out all you can about the ways it has been done before, and select the most appropriate approach. Next come quartermastering decisions—what kind of material, how much is necessary, and where can it be obtained? Then all supplies and equipment must be gathered. If need be, a prototype can be constructed to iron out the bugs, and you are on your way.

A Conceptual Approach

This book is based on the premise that media relate to one another in countless, valid ways, as long as their intrinsic differences are respected. One way to see this is to think of designs in a conceptual way. The charts that follow this section kaleidoscope conceptually; they detail the basic *ways of knowing* each medium. Many of our artists/craftspeople are combining media, mixing different materials successfully. This is not new; it is part of our cultural continuum. *No work is its own beginning or its own end.*

Below: The container-purse as a work of art—its subject matter and substance integrated with its origins—a raccoon. By William O. Huggins.

Right: The raccoon bag open with its flap out. In leather, fur, and bone, with ink drawings.

Below, right: Another view of the William O. Huggins raccoon bag, open.
Courtesy William O. Huggins

On Reading the Charts

Shapes are influenced by the working attributes of the material. Glass forms are rounded because blowing into a hot mass of liquid is like blowing up a balloon. The mass expands naturally into a round shape. This process/material interface influences the potential shape. Likewise, bronze is formed by alternately forging and hammering, and thereby raising a thick mass of hard material into thinner walls. Each material and process results in a predictable effect. The uniqueness of the container depends upon those personal aspects of a craftsperson's perception and expression.

Although not all processes outlined in the charts on pages 32–37 are described further in this book, there is a fairly good representation of different "ways of knowing."

When blown, glass naturally forms a globular shape much like a balloon. This large wine flask is from Corning Glass.

The mass of hot liquid expands naturally into a round shape in Kent F. Ipsen's globular glass vase (18" X 18"). *Courtesy Kent F. Ipsen*

The uniqueness of the container depends upon personal aspects of a craftperson's perception and expression. Helen Shirk takes quite a different approach to forming and constructing a copper container (6" X 7"). The surface was sanded with steel wool and has an oxidized finish. *Courtesy Helen Shirk*

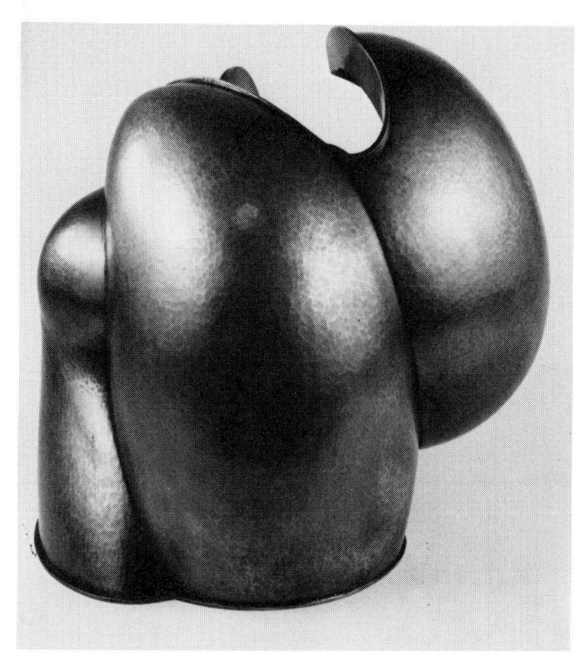

L. Brent Kington's free raised copper vase (9" high). Tool marks are used to texture the surface. *Courtesy L. Brent Kington*

MATERIAL	CHARACTERISTICS/ PROPERTIES	SIGNIFICANT PROCESSES FOR CONSTRUCTION/STRUCTURE	SIGNIFICANT PROCESSES FOR DECORATION—COLOR/PATTERN/TEXTURE	BASIC CONCEPTS FOR CREATING CONTAINER FORMS
CLAY earthenware stoneware porcelain	various kinds of consistency: loamy, pliable, sandy, gritty; various hardnesses when fired; various colors; leathery, liquid, pastelike	wedged, pinched, coiled, rolled, slabbed, draped, thrown on a wheel, cast in molds of plaster or sand, press-molded, raku; stoneware—porcelainware—earthenware—fired; mixed into thixotropic substance	colored with engobes, glazes, and slips of wide range of colors and textures; tooled, stamped or pressed, painted, dipped, sprayed, stenciled, resist-colored, sgraffitoed, slip-trailed, reduction fired, salt kiln fired	1. rolling clay into sheet, building by slabs, draping in mold, pressing into molds 2. coiling clay and building coil over coil, increasing and decreasing to form shape 3. turning on a wheel, raising and shaping through rotation of clay against pressure of fingers of both hands 4. diluting clay into a liquid (slip) and pouring into water-thirsty molds 5. pinching ball of clay into shape while rotating form on a wheel or in the hand
FABRICS natural animal vegetable synthetic	extremely wide range of textures, colors, and patterns due to origin and processes of original yarn, as well as weaving techniques and surface finishing; pressed, as for felt	cut, sewn by hand and machine, glued, laminated through appliquéing, patched, stuffed, gathered, folded, tied, pieced; formed by heat of thermoplasticlike acrylic	batiked, tied and dyed, painted, printed, embroidered, slump-worked, gathered, drawn-worked, combined with other materials	1. cutting, draping flexible sheet materials and attaching together with heat sensitive adhesive, sewing, pinning, or through use of mechanical fasteners 2. combining layers of material and stuffing batting between layers in various ways 3. cutting fabric into strips and plaiting or knotting
FIBERS natural unprocessed processed synthetic	wide range of colors and textures; wide range of thicknesses; wide range of elasticity, hardness, and softness even within one class, e.g., wool	netted, knotted, woven, macraméd, plaited, coiled, twined, knitted, crocheted, tatted, cut, tied	fringed, dyed, tied and dyed, e.g., ikat, combined in different color and pattern associations, combed, curled (if a thermoplastic synthetic, with heat), coated, painted, combined with other materials	1. using stringlike or wirelike lengths and forming by weaving in and out, or two lengths with various tensions around vertical stakes 2. plaiting strips by weaving multiple lengths in and out in various patterns

MATERIAL	CHARACTERISTICS/ PROPERTIES	SIGNIFICANT PROCESSES FOR CONSTRUCTION/STRUCTURE	SIGNIFICANT PROCESSES FOR DECORATION—COLOR/PATTERN/TEXTURE	BASIC CONCEPTS FOR CREATING CONTAINER FORMS
				3. coiling strips by wrapping and sewing one coil onto another 4. increasing or decreasing by looping with one or two needles; one loop into another (crochet or knit) 5. tying one length to another at various points with a variety of knots (macramé)
GLASS blown molded sheet	thermoplastic, transparent to opaque, full range of colors, fragile to strong, hard surface, chunks, pellets, sheet, blocks, canes, fluid properties optical imaging effects, matte to glossy surface	sagging, laminating, flame-worked, glass-blown, cast, fused, enameling, cut, joined with cames of lead or copper foil and solder, drilled, annealed, bent, bonded, draped	enameled, engraved, stained, painted, etched, stenciled, sgraffito worked, sandblasted	1. solid sheet melted into a mold 2. molten glass blown into molds 3. rods and tubes melted in flame while rotating and/or blowing 4. mosaic canes built up around female mold with male plug and fused until high temperatures are reached 5. pellets poured into cavity of mold left by melting out wax form and then melted 6. pieces of glass attached with lead cames or by wrapping edges with copper foil and soldering copper or lead together
LEATHER vegetable-tanned chrome-tanned	processed to be flexible, elastic, rigid, strong, thin, hard, soft, shiny, suedelike; various colors; various thicknesses	cut, bent, folded, creased, sewn, laced, punched, glued, riveted, hammered, plaited, woven, appliquéd, patch-worked, laminated, formed	dyed, batiked, painted, printed, tooled, embossed, stamped, carved, incised, burned, appliquéd, embroidered, fringed, combined with other materials such as cloutage work and lamelé	1. cutting and joining sheet material by gluing, stitching, using mechanical fasteners 2. wetting and forming in or over molds 3. weaving or plaiting strips 4. laminating by gluing sheets and then carving shape

MATERIAL	CHARACTERISTICS/ PROPERTIES	SIGNIFICANT PROCESSES FOR CONSTRUCTION/STRUCTURE	SIGNIFICANT PROCESSES FOR DECORATION— COLOR/PATTERN/TEXTURE	BASIC CONCEPTS FOR CREATING CONTAINER FORMS
METAL nonferrous base ferrous	opaque and crystalline structure, various melting points, various cooling points, fusible, various degrees of hardness, various degrees of ductility; various tensile strength, elasticity, brittleness, impact resistance; various wires, sheet, block, pellet, rod, tube	drawn, rolled, extruded, shotted, annealed, cut, drilled, carved, punched, soldered, pierced, molded, cast, repousséd, chased, split, raised, fused, filigreed, forged, welded, spun, turned, riveted, screwed, filed, hammered, crocheted (wire), tapped	engraved, embossed, etched, enameled, chased, filed, pickled, stripped, granulated, appliquéd or encrusted, stamped, damascened, niello-worked, filigreed, inlaid, braised, sanded, polished, electroplated, electroformed, burnished, ground, colored with chemicals and heat, hammered, combined with other materials	1. manipulating sheet materials in specific shapes and attaching by welding, soldering, or with mechanical fasteners 2. casting molten metals in molds 3. crocheting, knitting or weaving wire by increasing or decreasing loops 4. alternately hammering and heating (annealing), thinning and thickening sheet into shapes 5. filigree—coiling and bending wire in close configurations and soldering entire form into an open "solid"
PAPER AND PAPERBOARD newsprint printing papers fine papers coarse papers industrial papers tissue paper corrugated bristol poster railroad board	various colors, various weights, thicknesses, textures, from white to cream colored to gray colored, depending on material used to make the paper; coated and glazed, transparent, translucent, opaque, flocked, pressure-sensitive, smooth, rough, speckled	bent, crumpled, folded, braided, woven such as plaited, twined, coiled, fluted, wadded, twisted, punctured, torn, cut, curled, scored, used as pulp in papier-mâché and paper casting, sanded, glued, sewed	printed, painted, dyed, stained, textured by being folded, cut, crumpled, twisted, punctured, woven, overlaid	1. manipulating sheet material without decoration into specific shapes and attaching with glue, stitches, mechanical fasteners, self-fasteners 2. molding or casting pulp materials in/on negative or positive molds or on surfaces 3. twine or strip elements interwoven in a variety of techniques, building into a form element by element

MATERIAL	CHARACTERISTICS/ PROPERTIES	SIGNIFICANT PROCESSES FOR CONSTRUCTION/STRUCTURE	SIGNIFICANT PROCESSES FOR DECORATION— COLOR/PATTERN/TEXTURE	BASIC CONCEPTS FOR CREATING CONTAINER FORMS
PLASTICS acrylic (Plexiglas, Lucite, Acrylite) in sheet form	thermoplastic, transparent, translucent, opaque, water clear to a full range of colors thick as blocks to thin as sheeting optics imaging effects	softened and annealed, bent, blow molded, carved, cut, dulled, formed with heat in many ways, glued, planed, routed, sanded, tapped, laminated	buffed, colored with dyes, stains, paints, electroplated, engraved, etched, polished, sandblasted, silk-screened	1. sheet cut to various sizes and attached by gluing with solvent glues or occasionally with mechanical fasteners 2. sheet heated at point of bending until flexible and bent to shape with or without aid of jig 3. sheet completely heated until rubbery and molded through vacuum, blowing, or pressure in male or female forms
as paste	gesso and modeling paste, white to light gray	gesso, brushing, spatulating filling, modeling, can be built up	texturing and patterning, staining and rubbing off excess (antiquing)	1. applied in a buttering process to a matrix
PLASTICS epoxy as liquid	pourable, translucent to transparent, water clear to straw and dark amber when filled, cream colored to black or metallic hues, thermosetting plastic	draped when impregnating fiberglass, coated, cast, cut, drilled, glued, laminated impregnators, molded, planed, poured, routed, spread with a spatula, tapped, used to contain embedments, used as an adhesive	buffed, colored, engraved, polished, sanded, coated	1. cast in a mold 2. impregnated into fiberglass and draped or wrapped in a male or female mold 3. used as a coating over a positive form 4. solid pieces (poured sheet that cured) cut and then adhered at joints with more resin 5. poured as matrix between/around stained glass chunks
as paste	pastelike viscosity	spread with a spatula, essentially for gluing and coating	as coating can be colored, can be used to fill holes in wood to become decorative aspect of wood surface	

MATERIAL	CHARACTERISTICS/ PROPERTIES	SIGNIFICANT PROCESSES FOR CONSTRUCTION/STRUCTURE	SIGNIFICANT PROCESSES FOR DECORATION— COLOR/PATTERN/TEXTURE	BASIC CONCEPTS FOR CREATING CONTAINER FORMS
PLASTICS foams (Styrofoam polyurethane foam, etc.)	various densities as rigid and as flexible blocks and sheets; opaque; thermoplastic	built up block by block or piece by piece, carved, coated, drilled, foamed in place with a two-component system, glued, heat-cut, physically cut, vaporized as lost-foam casting; as mold for casting other materials such as cement or epoxy; as maquettes for other materials	coated, textured with solvent, painted	1. foamed in place in a mold or over a base form 2. solid units built up by gluing 3. solid units carved mechanically or with a knife 4. Styrofoam casting—Styrofoaming positive encased in a mold, vaporized with heat, leaving a cavity for pouring of metal or plastic
PLASTICS fusible Dec-Ets Poly-Mosaics (polystyrene, acrylic, and other thermoplastics)	pellets in all colors from transparent to opaque; mosaic tiles (Poly-Mosaics) in stained glass colors	after fusing with heat or solvents: bent, cut, drilled, glued, molded, sagged, slumped, laminated	buffed, polished, sanded, overlapped, and re-fused	1. pellets or mosaics fused on flat sheet with moderate heat and while hot bent into a form 2. hot sheet sagged into or draped over a mold 3. fused units adhered to one another at joints with solvent glue 4. sheets laminated with heat or solvent and cut into shape
polycarbonates	thermoplastic similar to acrylic—see acrylic	see acrylic	see acrylic	same as for acrylic
PLASTICS polyester resin	thermosetting resin, as liquid and paste when filled with a thixotropic filler, transparent and water clear to opaque	coated, cast, cut, draped (when impregnating fiberglass), drilled, filled, glued, impregnated, laminated, laid-up with fiberglass, planed, poured, routed, sprayed, spread with a spatula, tapped, used as an adhesive, used to contain embedments	buffed, colored, engraved, polished, sanded	same as for epoxy

MATERIAL	CHARACTERISTICS/ PROPERTIES	SIGNIFICANT PROCESSES FOR CONSTRUCTION/STRUCTURE	SIGNIFICANT PROCESSES FOR DECORATION—COLOR/PATTERN/TEXTURE	BASIC CONCEPTS FOR CREATING CONTAINER FORMS
silicone	thermosetting, translucent—pastelike material to opaque, pourable liquid, milky clear to opaque paste	cast, dipped, molded (used as mold material), spread with spatula, squeezed into patterns as a paste	colored	1. made into female molds 2. as paste used as adhesive or as matrix
vinyls	thermoplastic milky liquid that dries water clear to pastelike material and in sheet form	carved, cut, coated, dipped, glued, impregnated, molded, sewed, sprayed, used as an adhesive (Sobo, Elmer's, etc.)	colored, sewed, and stuffed	1. used as an adhesive 2. sheet used for fabric, sewed into shapes 3. sometimes used to impregnate fabric and then shaped by draping in a mold or wrapping around an armature or mold
WOOD softwood hardwood processed wood (over 100,000 different varieties)	no absolutes, soft to hard, varying degrees of shrinkage; when seasoned, varying degrees of resistance to decay and insects; bendable, variations in grain, color, and figure; when cut, cut in different ways and thicknesses from paper-thin veneers to thick sections; various qualities of toughness, resistance to warpage	cut, planed, carved, drilled, chiseled, joined, jointed, chamfered, tapped, gouged, filed, rasped, glued, whittled, laminated, bent, turned, assembled, sanded, vacuum-formed, attached with mechanical fasteners	sanded, oiled, painted, stained, antiqued, carved, incised, textured with heat and flame, sandblasted, veneered, inlaid, marquetry-worked	1. elements of flat pieces are built by joining through use of joints and/or glue, or mechanical fasteners into forms 2. solid blocks are carved or cut away by gouging out unwanted areas while the block is rapidly rotated 3. small, flat pieces are laminated to one another using glue (sometimes with dowels) and pressure in or around a male or female mold, drying into new, larger shape 4. boards are steamed or boiled in hot water and bent by being clamped to a mold and allowed to dry into new configuration 5. solid blocks are carved by hand or with portable hand tools

A perception, in porcelain, by designer Enzo Mari for Danese of Milan, Italy. *Courtesy Enzo Mari and Danese, Milano; photo by Jacqueline Vodoz*

A winter handbag woven (and crocheted) so that a hand can fit into the glove while holding the bag. By R. A. T. Stevens. *Courtesy R. A. T. Stevens; collection Ms. Helen Bangs*

BASKETRY 3

Containers of natural fibers have existed in various forms for centuries all over the world. And probably because of the consistency of available materials, basketry techniques and design have shown great continuity. Although one sometimes can clearly identify regional styles, plaited and coiled baskets from Africa, the Americas, and Asia can be extraordinarily similar. That does not mean that contemporary craftsmen come to basketry with no room for creativity. Variations stem from new applications of old materials, and utilization of media like wire, newspaper, and seaweed. Striking new designs combine ancient techniques, unusual materials, and contemporary design idioms.

There is a universality of shape and design to the basket arts. This coiled form of grass bound with raffia is from Kigezi, Uganda, Africa. Weaving is done by pastoral Watusi women.

Above: The raffia coiled basket on the lower right is from Sierra Leone, Africa. The one on the left is raffia with colored cotton designs and is from Guatemala. The basket at the top is from Mexico.

Right: A plaited basket from Japan.

Bottom, right: Two miniatures that replicate the real thing from Thailand.

Below: An unusual twined basket of wire (about 18" tall) from Luzon, Philippines.

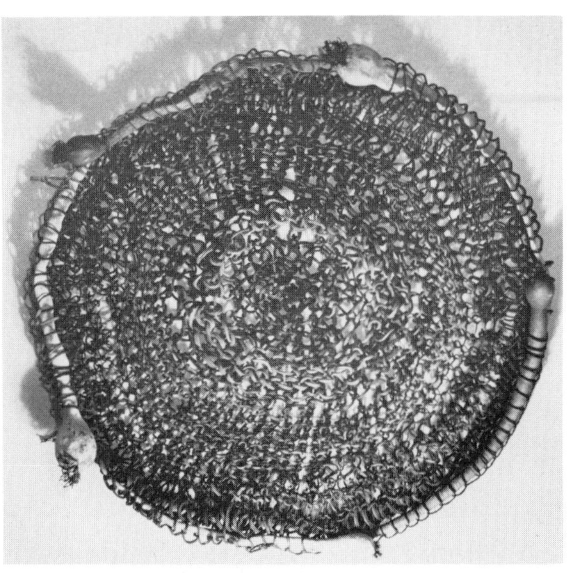

Top, left: An atypical use of materials. This basket was twined of newspaper. By Ed Rossbach. *Courtesy Ed Rossbach*

Top, right: Lucy Traber created this netted basket of seaweed.

Center, left: A coiled basket by Joan Sterrenburg in a contemporary idiom. *Courtesy Joan Sterrenburg*

Center, right: A sewn basket of plant husks from Japan.

Bottom: Julie Connell's wool-yarn-wrapped sisal basket.

Raffia in a partially woven, partially embroidered pattern strung on a wire armature, from Colombia, South America.

A creative birthday cake basket coiled in raffia by Sally Edmondson.

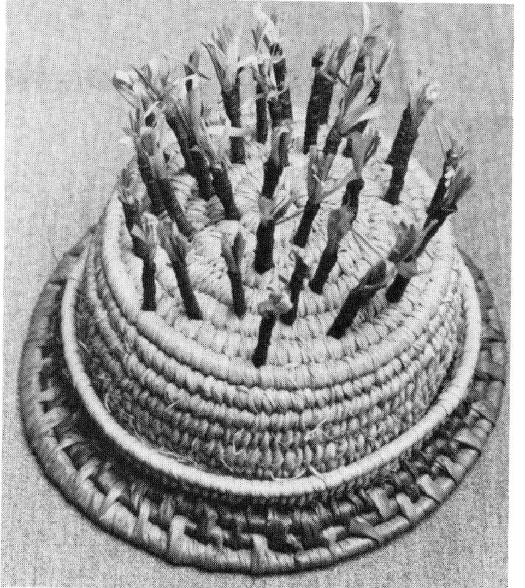

Materials

Although basketmakers in industrial societies often harvest their materials from catalogs, many satisfying substances can be found in any environment. Palm fibers, bamboo, grasses, raffia, split wood, roots, stems, and leaves are frequently used to create baskets. Containers of wire, paper, seaweed, yarn, and hair are more novel. But since the material becomes structure, surface, color, and design in a basket, a variety of tactile and visual qualities is essential to the basketmaker.

In addition to the many styles of weaving, wrapping, coiling, and plaiting, two other categories of processes figure in the basketmaking process: preparing the material for construction and dyeing or coloring the fiber.

Some fibers are formed while green and are easy to bend and twist. Stiffer, harder, or dryer fibers are usually soaked to make them soft and pliable.

Grasses and leaves are dyed by soaking or boiling them with barks, fruits, or other color-yielding materials. Onion skins, ash, oak, charcoal, nut shells, tea leaves, and indigo are just a few natural dyes. The most successful colors tend to be subdued golds and browns. Aniline dyes and harsh fluorescent colors, which have gained prominence because of their availability, do not seem compatible with the natural quality of vegetable fibers; they scream out in contradiction.

Availability and Special Preparation of Materials

Bamboo can be found growing on the southern highlands (Tennessee, Kentucky), where it has been used by the Cherokee Indians and mountain basketmakers. Bamboo is hard and tough and requires soaking to render it weavable.

Splints (upright spokes of baskets) have been formed of hickory, oak, birch, linden, and willow. Pieces are cut into one-eighth- to one-half-inch

Palm leaf plucked from the ground is used for basketmaking. This scene is from Honduras, Central America.

The materials of baskets can also be harvested from catalogs as is this synthetic raffia coiled over sea grass in this basket "Mickey Mouse" by Ed Rossbach. *Courtesy Ed Rossbach*

widths. By beating a log with a wooden mallet, pieces can be stripped with a knife or heavy shears as the fibers separate.

Willow tips and shoots, cut in the spring or early fall, are flexible enough to weave. Cattail or rushes (varieties of *scirpus*), gathered before they mature, are excellent weaving materials when dried slowly. The Hopi Indians of the Southwest use all parts of the Yucca for spokes and for wrapping coils. Varieties of the sumac (*Rhus tubobata*), known as rabbit bush, are cut in summer, soaked, and debarked. Then the sapwood is cut into thin strips and wrapped around the rest of the twig in the coiling process. Wild honeysuckle vine should be cut in the winter, wound into rolls, boiled for four hours, soaked overnight, rinsed, and dried in the sun.

Shedding bark of the eucalyptus tree is also soaked. Dracena palm leaf, which grows in cool, coastal areas of the West, is worked wet. Grasses of various sorts are dried in the shade slowly. Seaweed (*nereocystis*), or kelp, found washed up along Pacific coast beaches, is dried, soaked in hot water, and woven when wet.

Reed and raffia usually are purchased from suppliers who import these materials.

Yarns, threads, wire, and newspaper, gathered from their usual sources, need no special preparation.

Tools

All these materials must be flexible to be woven. If they are not, they require soaking. A water supply in a tub or basin should be available.

The only other necessary supplies are shears or knives for cutting the materials into proper working lengths; an awl; a knitting or darning needle is handy to poke holes or for sewing parts together. Sometimes a pair of pliers is useful to help pull through a difficult length of fiber or to yank out a needle that has gotten stuck.

Basketmaking Process

Coiling, twining, weaving, plaiting, and netting are the essential basketmaking processes. Sometimes elements are embroidered, sewn, or glued onto the surface for special effects.

COILED BASKET (AMERICAN INDIAN STYLE) by Sally Edmondson

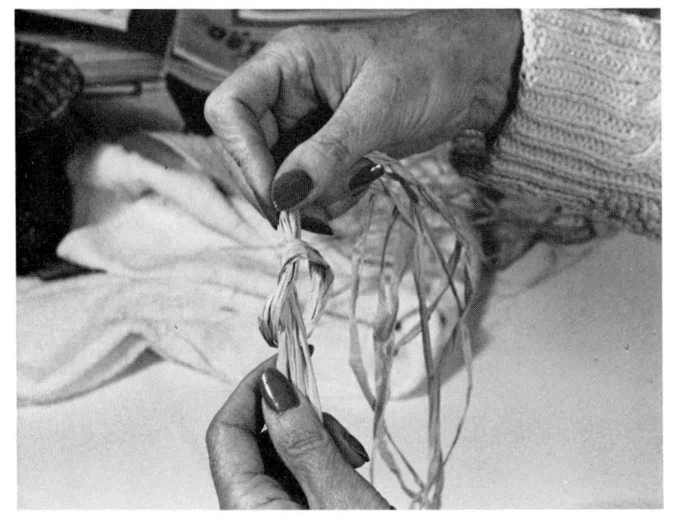

An overhand knot is tied in the center of six strands of raffia.

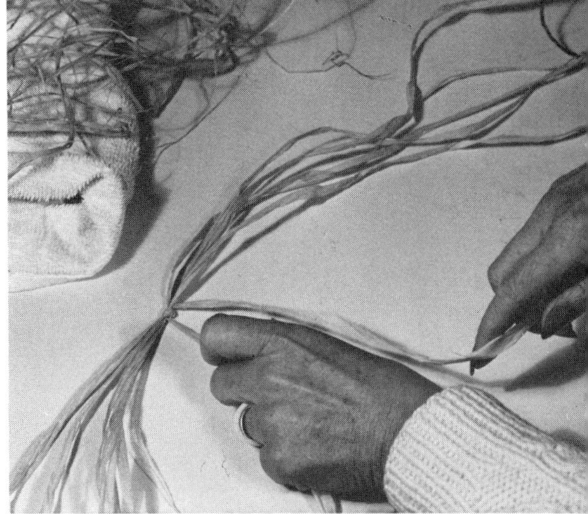

Each strand becomes a spoke and is separated and knotted one to the other on the left and on the right for all twelve spokes.

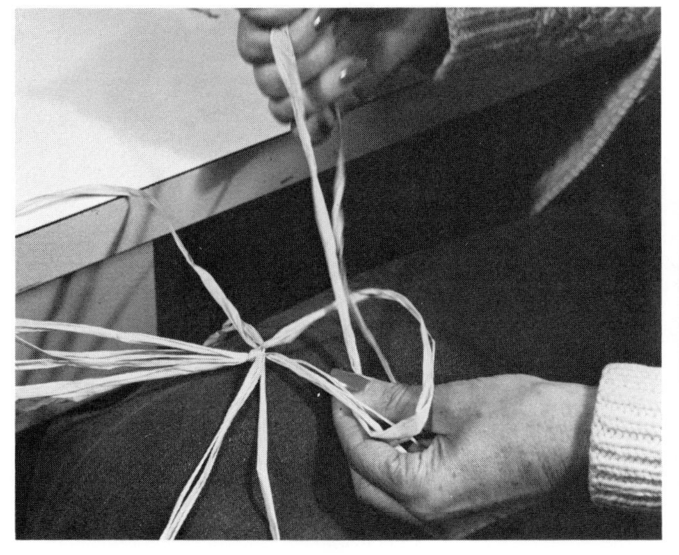

Each strand then is knotted twice all around the center knot.

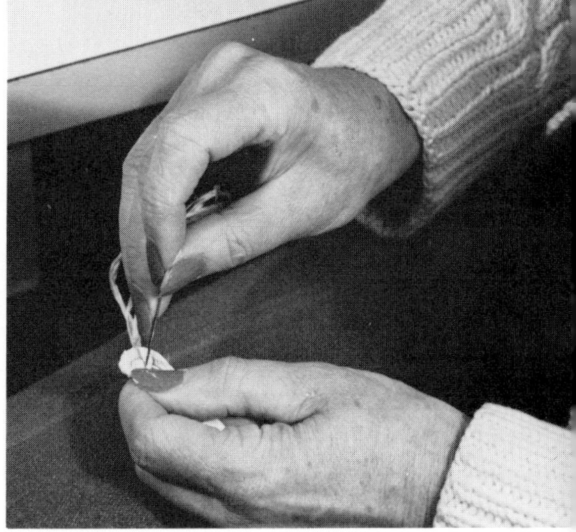

The longest strand is found and is threaded into a needle. (American Indians fashioned needles out of bone.) This becomes the stitching spoke. All strands are gathered to make a core, and an overhand stitch is sewn between each knot around the center of the coil.

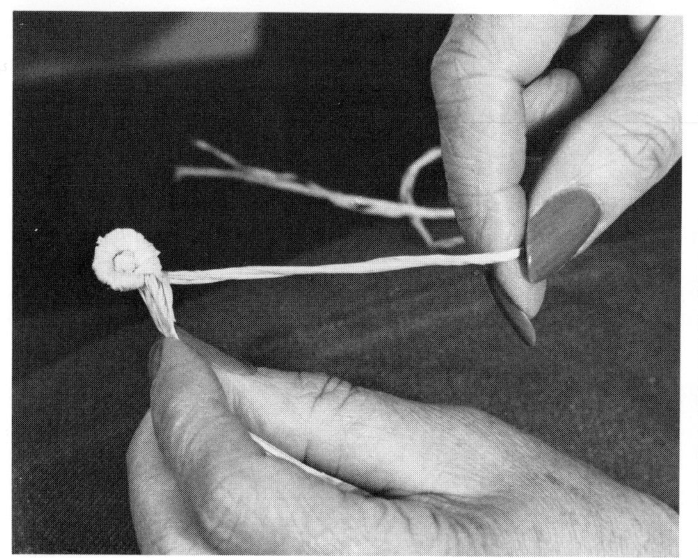

When a thread is used up, the next longest strand is threaded and used.

New strands are added as the old ones shorten . . .

. . . and the longest strand is used for stitching around each coil in a figure 8 into the juxtaposed coil.

A completed basket by Sally Edmondson with crow feathers below the rim.

Another basket coiled with wool yarn by Sally Edmondson.

Cording is wrapped with Grosheen fiber and anchored in standard coiling process by Jan Hoffmann. Loose fibers are tied on with a hitch called a lark's head knot (rya). Two ends of the string are passed through the loop (*right*), which is wrapped around the coil.

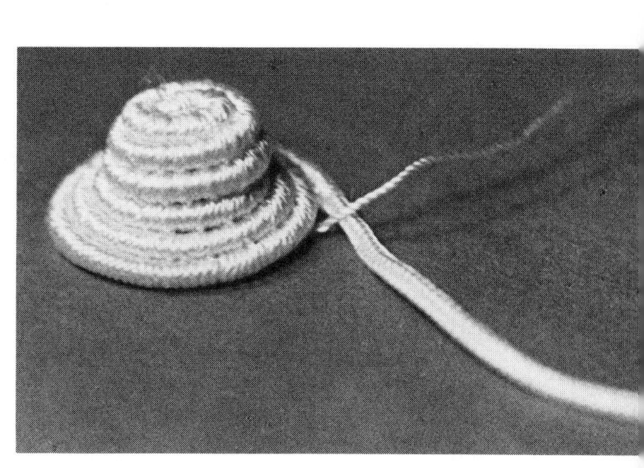

The lid is coiled in a similar fashion until it matches the opening in the base.

The pot in lined with silk satin.

Jan Hoffmann's completed pot is just a few inches tall.
Photos Courtesy Jan Hoffmann

Lida Gordon uses counterchanged colors in a coiled linen basket. *Courtesy Lida Gordon*

"Peppermint Eden" by Rebecca A. T. Stevens. *Courtesy R. A. T. Stevens*

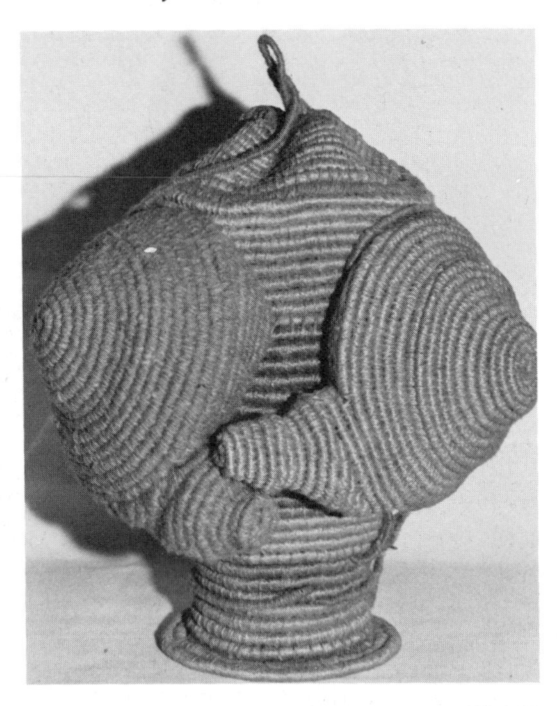

"Mende" by Rebecca A. T. Stevens. *Courtesy R. A. T. Stevens*

"Soul Catcher" (9" X 9") by Louise Robbins. A feather-lined basket of wool and reed. *Courtesy Louise Robbins*

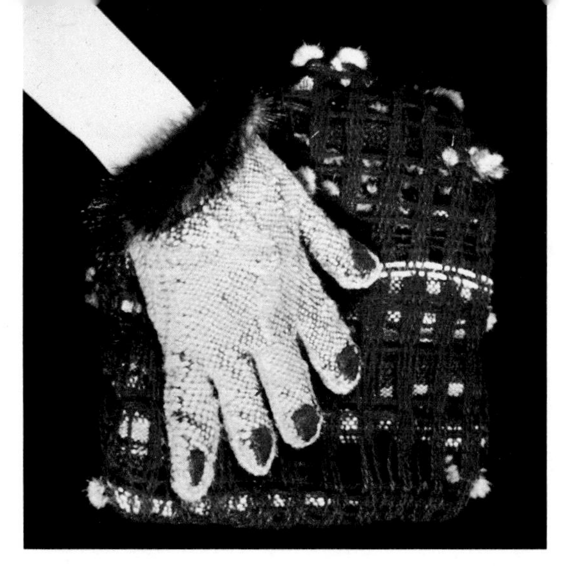

Winter Handbag, woven and sewn natural fibers and fur by Rebecca A. T. Stevens. *Courtesy Rebecca A. T. Stevens*

A burl bowl by Mark Lindquist. *Courtesy Mark Lindquist*

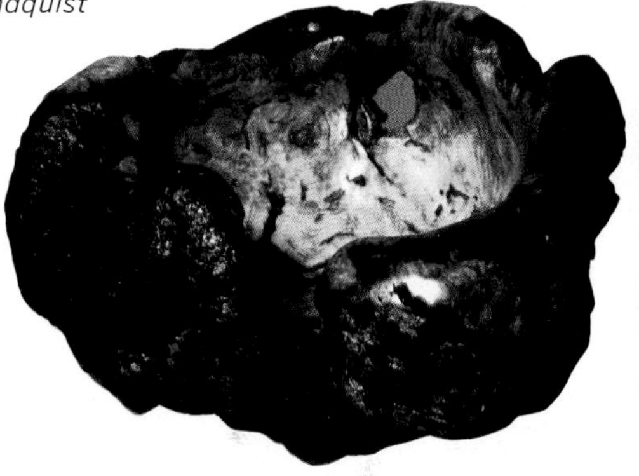

A plaited ribbon basket by Susan Jamart.

A stoneware mug with personal identity by Dave Boronda. *Courtesy Dave Boronda*

Constructed leather box by Marc Goldring. *Courtesy Marc Goldring*

Copper Form I (6" x 6"), formed and fabricated copper by Helen Shirk. *Courtesy Helen Shirk*

Gourds lined with beads embedded in beeswax, fetish figures from Mexico.

Horse Heaven, crochet and cloth by Norma Minkowitz. *Courtesy Norma Minkowitz*

Cactus Basket (7" high), crocheted cotton by Renie Breskin Adams. Cactus and sand form a removable lid. *Courtesy Renie Breskin Adams*

Sas Colby's Autobiographical Box (5" x 8" x 8") is silk over cardboard and stuffed with Dacron filler. The dolls have photo-transfer faces of members of her family. *Courtesy Sas Colby*

Mace of fabricated metal and walnut by Harlan Butt. *Courtesy Harlan Butt*

Container (10" x 5½" x 2") of cast acrylic and electroformed sterling silver by Daniella Kerner. *Courtesy Daniella Kerner*

David Luck's cast-iron container. *Photo by David Luck*

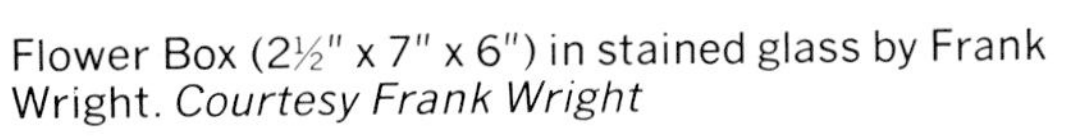

Flower Box (2½" x 7" x 6") in stained glass by Frank Wright. *Courtesy Frank Wright*

Lidded container of reduction stoneware (28" high) with platinum luster by Greer Farris. *Courtesy Greer Farris*

Michael Riegel's Purse (17" long) of fabricated steel and leather with surface drawing. *Courtesy Michael Riegel*

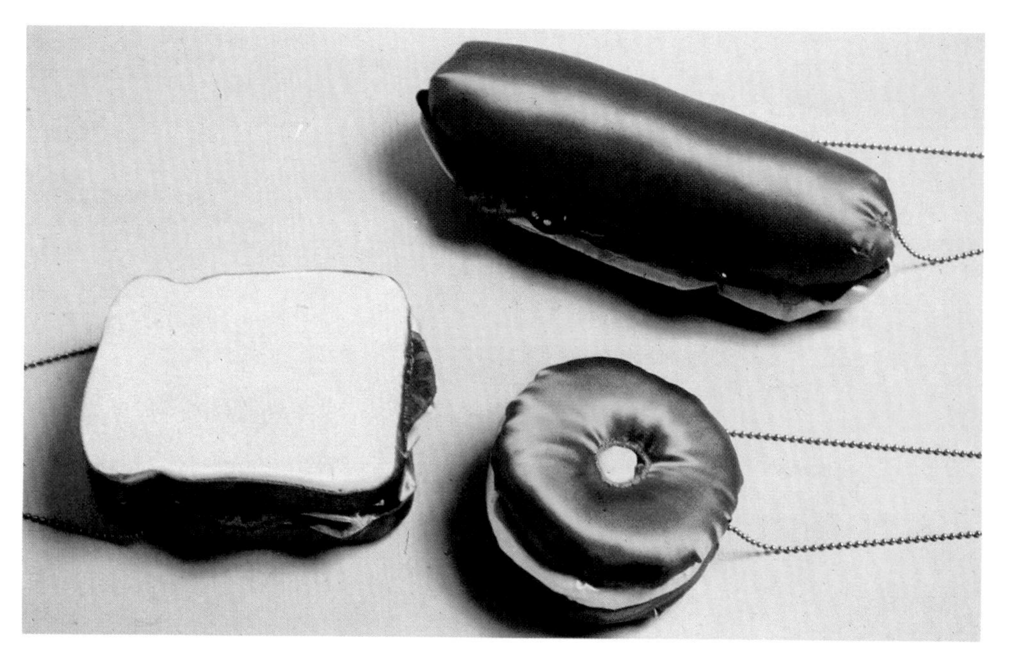

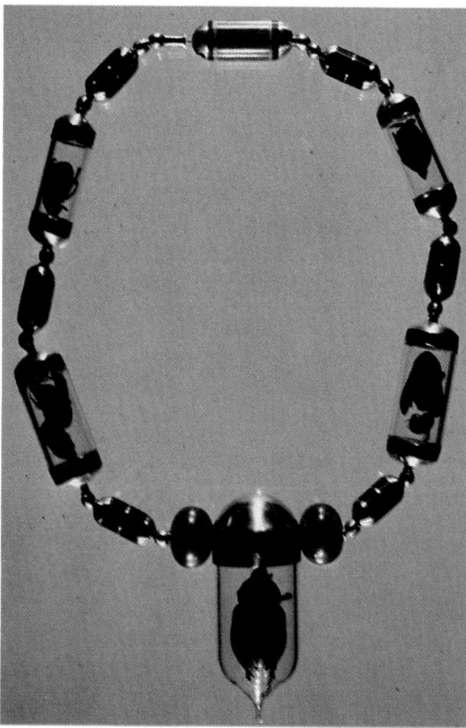

Satin sandwich bags by Joy Nagy. *Photo by Bob Hanson. Courtesy American Crafts Council*

Entomology Necklace #1 by Jan Brooks Loyd. *Courtesy Jan Brooks Loyd*

Pig in a Poke by Jim Cotter. Leather, brass, and cast sterling silver. *Courtesy Jim Cotter*

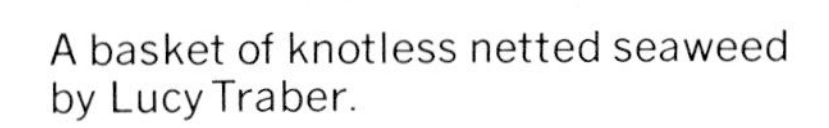

A basket of knotless netted seaweed by Lucy Traber.

An electroformed copper container and lid inlaid with 24-karat-gold and silver foils and epoxy resin by Barbara Anne Nilaũsen. *Courtesy Barbara Anne Nilaũsen*

Glazed ceramic box by Florence Cohen. *Courtesy Florence Cohen*

Victorian Rose (3½" high), coiled basket with quilted lining by Jan Hoffmann. *Courtesy Jan Hoffmann*

Lisa Drumm's whale doll contains another doll. The whale's mouth zips open and shut. *Courtesy Lisa Drumm*

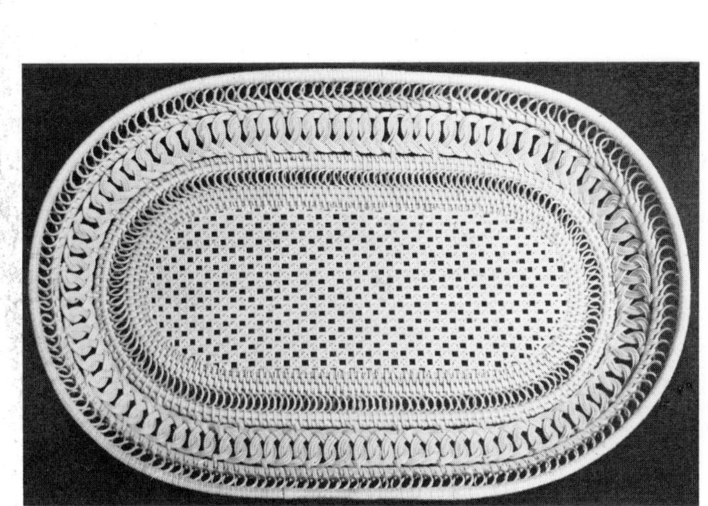

A combination of coiling and weaving in a rattan basket tray from Japan.

Coiling

Coiling is one of the easiest and earliest methods of basketmaking. It is similar to forming coiled pottery. Stiff bundles of fiber are wrapped, coiled into a spiral, and stitched together in a variety of ways. Often the method of stitching one layer of the spiral to another creates a decorative effect.

Some kind of pointed instrument is used to spread a hole so that a soft grass, leaf, or some other fiber can be inserted to stitch and bind one coil to the other.

Stiffer coiling materials often require soaking to make these pliable, and if different colors are to be incorporated in the binding and/or wrapping fibers, these require dyeing beforehand.

Shapes and sizes can be determined by controlling the circumference of each coil. If it is allowed to grow larger in circumference than the previous coil, the shape will expand; conversely, smaller coils will reduce the size.

COILING WITH EPOXY

Instead of connecting coiled surfaces by stitching, a rapid curing two-part epoxy is used. The two components are mixed together, applied to the points of joining, and held until cured.

Some kind of mold form is necessary. In this case it is a large glass container, which was waxed first so that any oozing epoxy would not attach itself (and the hemp cord) to the glass.

After the coiling has been completed, the end is tapered by removing bits of hemp with an X-acto knife.

The mold container is removed.

A sturdy hemp form. Changing the diameter of the hemp and the size of the form could vary the function of this container from a pencil holder to a wastepaper basket.

COILING BY WRAPPING—Julie Connell

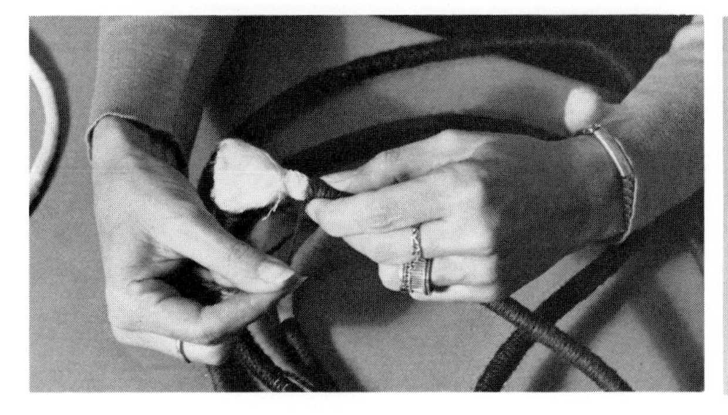

Julie Connell creates a range of shapes by varying the base, or stuffing (such as upholstery cording, which comes in ¼" to 1" diameters), and the wool yarn, which is wrapped tightly around the cording.

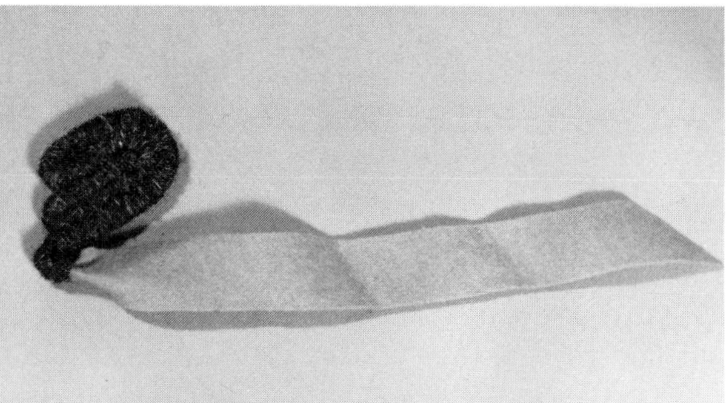

Here Julie Connell uses felt strips as her base. Coiling can be random as well as symmetrical.

Chenille yarn was used to wrap upholstery cording, then coiled. Juxtaposed parts are stitched together with needle and yarn.

The completed bowl by Julie Connell.

Coiled and wrapped yarn basket, 8" tall, by Julie Connell.

COILING BY THE BUGINESE OF SULAWESI, INDONESIA

In forming a basket Buginese style, the dried center of a palm leaf is moistened and wrapped around a cylinder to train it to curve in the proper direction.

The dried palm leaf itself is prepared and sliced into the proper width, using a razor blade.

The coil is rolled to the proper diameter and set into the previous coil. (Wider diameters create wider openings.)

Using flexible palm leaf, a figure 8 coiling stitch is used to attach one coil to another.

Three completed Buginese lidded baskets.

Knotless Netting

Netting as a basketmaking process has been adapted from the making of fishing nets. Essentially, the technique involves looping one length into another. The shape can be controlled by looping more than one loop in a previous hole, or skipping a loop to reduce a circumference. Kelp baskets are made this way in a manner very similar to that used by the Eskimos.

NETTED SEAWEED BASKETS by Lucy Traber

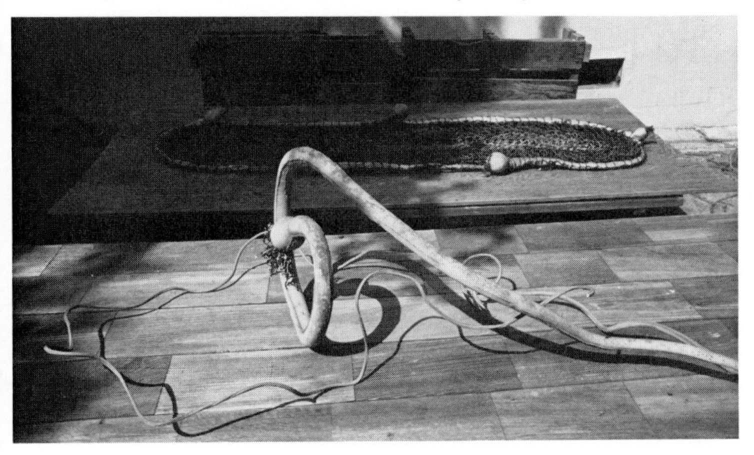

Nereocystis (Phaeophyta), commonly called kelp, is gathered from a southern California beach and allowed to dry in the sun. A discontinuous supply washes ashore in California, Alaska, and the Aleutian Islands.

It is soaked for several hours in hot water and woven while wet.

An initial loop is formed . . .

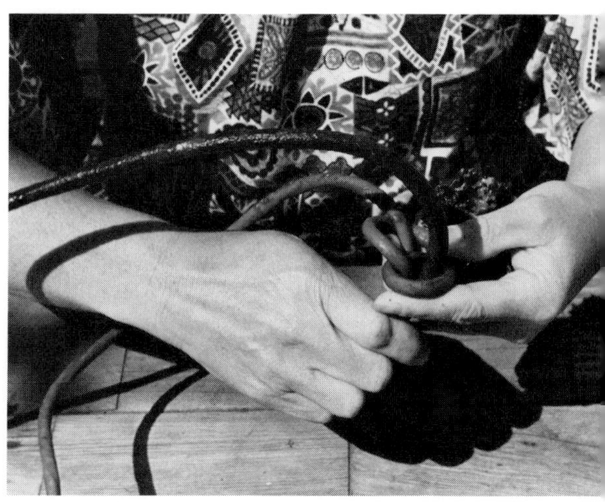

. . . and successive loops are made from the first one, using a buttonhole stitch.

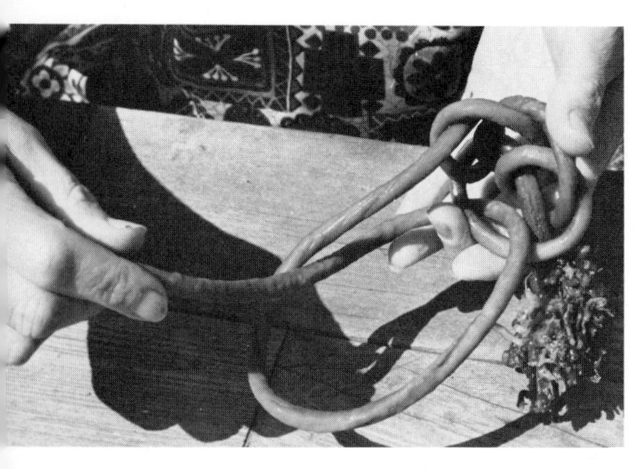

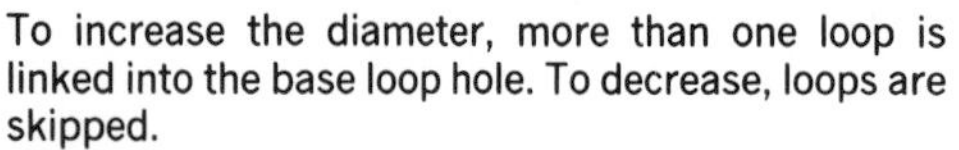

To increase the diameter, more than one loop is linked into the base loop hole. To decrease, loops are skipped.

Lucy Traber pulls the kelp tightly—or loosely—to help control the shape. As it dries, the kelp wiggles and squirms into a tight and textured weave.

Two loops are made into a previous one. No matter how tightly the kelp is pulled, some gaps are to be expected. It is the nature of the material. While wet, it is soft and almost slippery; it will not stand up at this point and is much like chain mail . . .

. . . but it dries into a curly, woody, hard texture. The gourd-like forms help to keep the kelp afloat when in the water. Roots anchor it to the floor at the other end. The sea otter ties himself onto the kelp when sleeping to stay afloat and avoid floating away. A completed basket by Lucy Traber.

Another kelp basket, attached to driftwood, by Lucy Traber. Kelp, by the way, should *not* be plucked from the sea and dried because it does not dry free of mold when picked fresh. It takes three days of hot sunshine to dry kelp that washes up onto the shore.

Plaiting

Plaiting, weaving without a loom, is probably the most popular process for making containers. A wide range of shapes and designs is possible with plaiting. And a wide range of materials is available—from pandanus, palm, and grasses to paper and ribbon. Tools are simple—only a scissors or knife is necessary. The results are serviceable and handsome. Beautiful and intricate patterns emerge when different colors are used. Shapes vary, as well, from rectangular boxes with lids to envelope shapes, carrying baskets borne on one's back, trays, and ceremonial offering baskets.

Usually plaiting results in a flexible, yielding material that gives when weight is pressed against the plaited surface; therefore, a support system is often necessary. The only exception to this is when a stiffer, more rigid, fiber is used. When support is necessary, a decorative skeletal structure of a stiffer material may be lashed to the outside or inside of the basket. For some purposes mounting the basket on a rigid base is sufficient. Or, lining the plaited form with another reinforcing layer of the same plaited material often can maintain a rigid structure. Similarly, where a lid slides over a base, that second layer provides a bit of necessary support. Unlike coiled, twined, or woven baskets, plaited forms have to be designed to integrate the necessary reinforcement.

Pattern is created through the interweaving of different colored elements. Plaited baskets employing wide elements usually sport simple patterns. To achieve intricate patterns, the weaving elements have to be fine—narrow enough for a pattern to emerge, yet proportioned to the size of the container. (The smaller the container, the finer the elements should be.)

The type of design is determined by the width of the wefts, the system of plaiting, and the number of colors introduced.

Basic plaiting is weaving of elements under and over one another at 90° angles (even at corners). In a plain weave, elements are woven one over and one under. More complex designs are created by weaving under or over two or more, and grouping these variations in a particular repeating pattern. As a basket is plaited, elements end up spiraling around the form until they terminate at the top. In addition to two-weft plaiting, there can be three or more weft plaiting to achieve other designs.

WEFT PLAITING by Susan Jamart

1 Set out elements and plain weave, one over and one under.

2 At the beginning, all elements are plaited at 90° angles. New elements are introduced until the center woven area approximates the desired size of the base.

3 Elements are temporarily fixed with masking tape.

4 To economize on material (and to create a symmetrical basket), set out the side elements in steps as shown here. If an asymmetrical basket is desired, this is not necessary.

5 These strips are placed here by Susan Jamart only to graphically demonstrate how the basket is to be folded.

At the corners of the temporary paper delineators, the basket is folded. 6

7 The trick is to cross the first two elements. After the initial weaving, elements will want to move in diagonals.

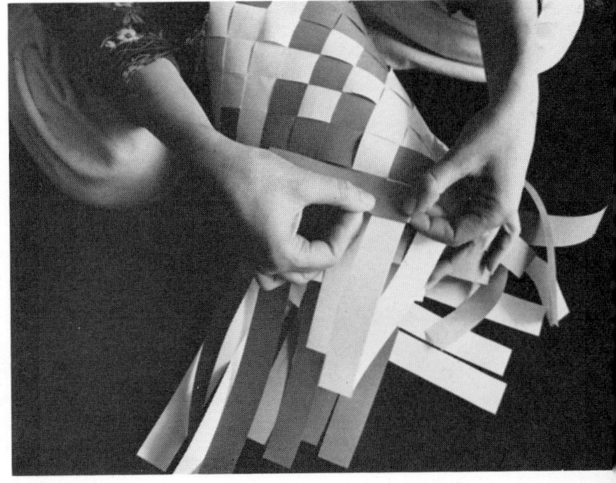

Just plait them in the diagonal direction. If different colors are used, it is easier to differentiate among the elements. 8

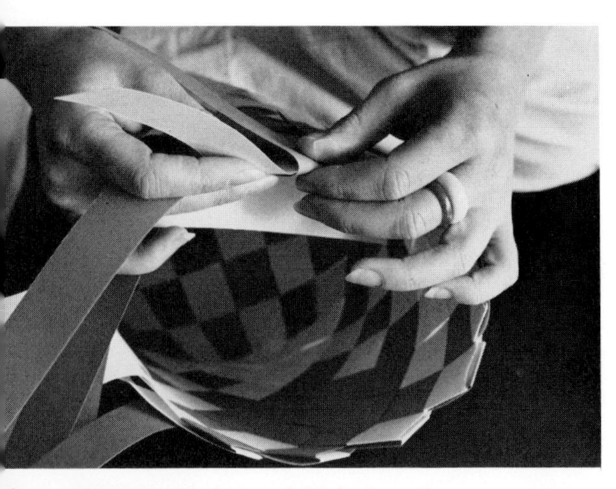

9 Elements spiral around the form as they are woven one under, one over.

10 One way to finish the top is to form sawtooth elements as shown here. To effect a closure, plait elements back into the basket.

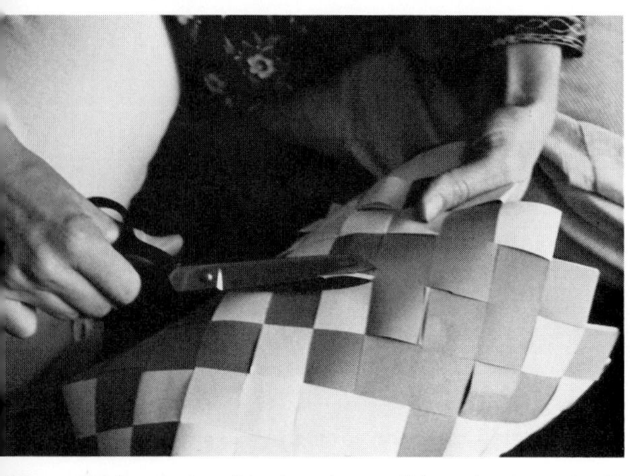

11 Fold each strand back and weave it in and out over itself. Trim excess with a scissors.

Susan Jamart's completed two-weft paper basket.

Colors can be controlled, as in Susan Jamart's three-weft plaited ribbon container, by preplanning the positioning of colors.

These are birchbark baskets from Finland that have been two-weft plaited exactly as demonstrated by Susan Jamart, except that the edges are finished straight (folded in diagonals) along the rims rather than sawtoothed.

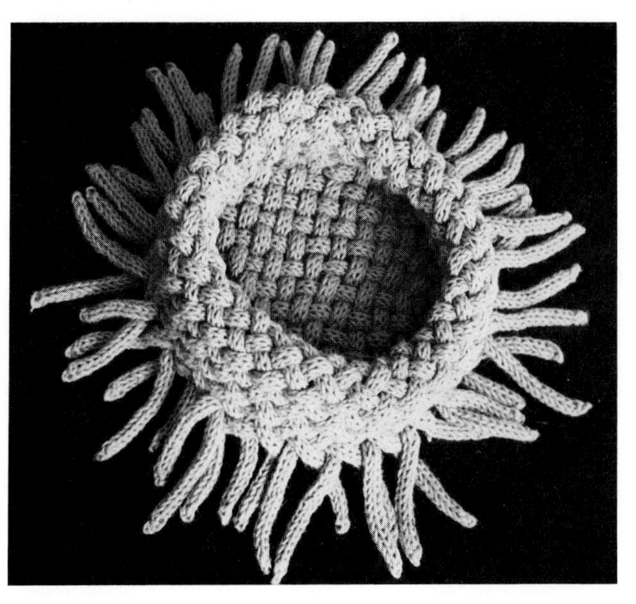

Horse-rein elements have been two-weft plaited in a cylinder that folds back on itself, with ends resting on the table. By Susan Jamart.

Twined and Woven Baskets

Twined and woven baskets have a few points in common—elements are woven in and out of a spoke system, and usually they are more rigid than plaited forms. These two processes vary inasmuch as twining involves the interweaving of two elements at the same time around spokes, whereas weaving is simply weaving one element in and out around spokes.

Design variations occur as colors are changed and elements are skipped or woven in diagonals or other patterns.

Procedure for Creating Baskets

1. Preparation of grasses or fibers
   a. cutting
   b. drying
   c. dyeing
2. Soaking fibers (when needed)
3. Interweaving elements: coiling, netting, plaiting, twining, or weaving
4. Ending elements
5. Trimming, finishing ends

SIMPLE WEAVING

Dried palm spines are split and prepared by soaking.

Spokes are aligned on the ground and an initial strip is woven in and out of the spokes to set them in place.

Weaving continues to form the base, with elements pushed close together so there are no gaps.

After the desired circumference has been achieved, the spokes are bent upward . . .

. . . and the weaving continues to form the basket sides. New weavers are introduced by overlapping them with the ends of the previous weaver.

The top of the basket is completed by weaving the weaver in and out around the top after spokes have been trimmed.

Ends are finally cut and buried.

The completed basket.

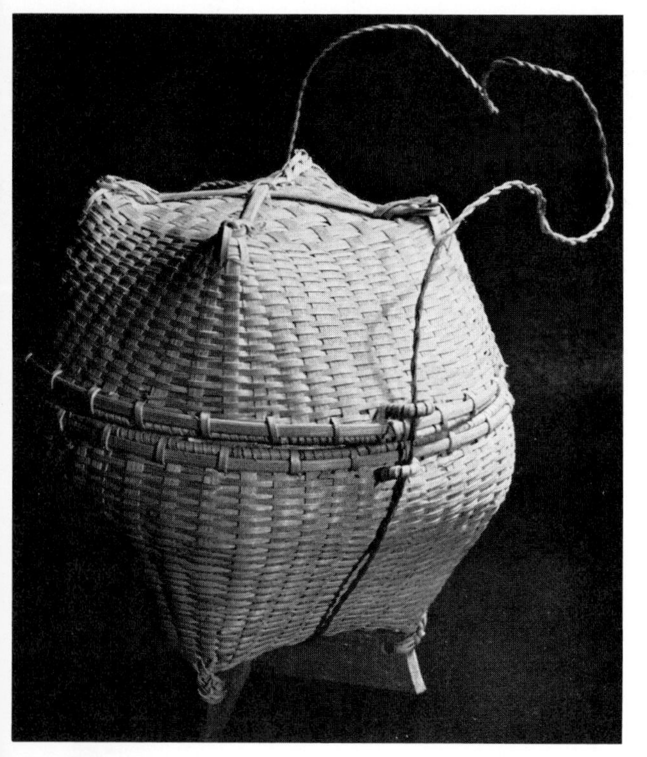

Woven basket from Chiang Rai, Thailand.

WEAVING WITH WIRE

Spokes are formed by hammering wires flat and arranging them in a wheel-like arrangement. One spoke is turned back (to create an odd number) after fine wire has been used to bind the spokes together.

Wire is woven in and out, controlling the diameter by tightening or releasing the weavers.

When the form has been woven to the proper height, ends are bent in, trimmed with a wire cutter . . .

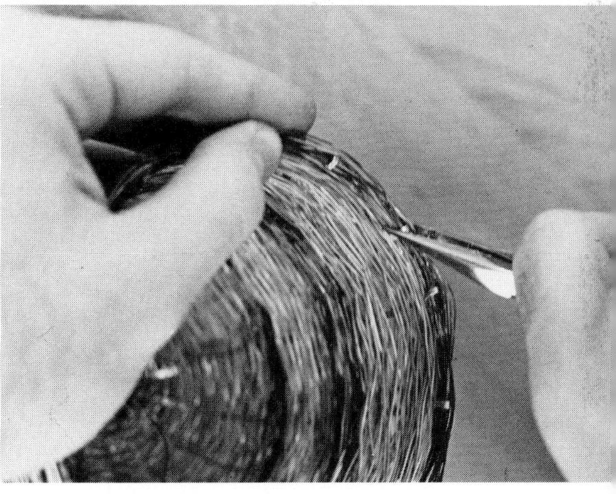

. . . and tucked back into the weaving using a pliers.

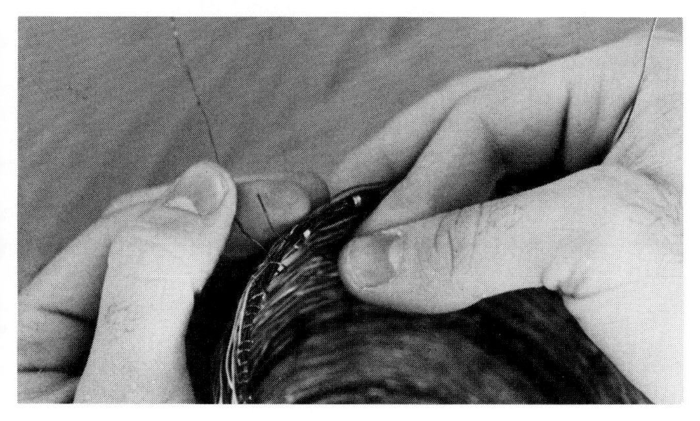

Fine wire is then whipped around the top to finish the basket.

The completed basket. Different kinds and colors of wire—brass and copper—help to create a striped pattern.

A brass basket box from India. The spokes are soldered onto a solid brass base. The top is started through a brass "life-saver" shape.

Right: Top rims around base and lid are soldered in place, as is the hinge.

A woven bottle holder made of Kraft cord. Spokes are woven back into the basket and then knotted to create a decorative effect.

Dracena palm leaf that grows in cool coastal areas is worked wet around willow branches. By Lucy Traber.

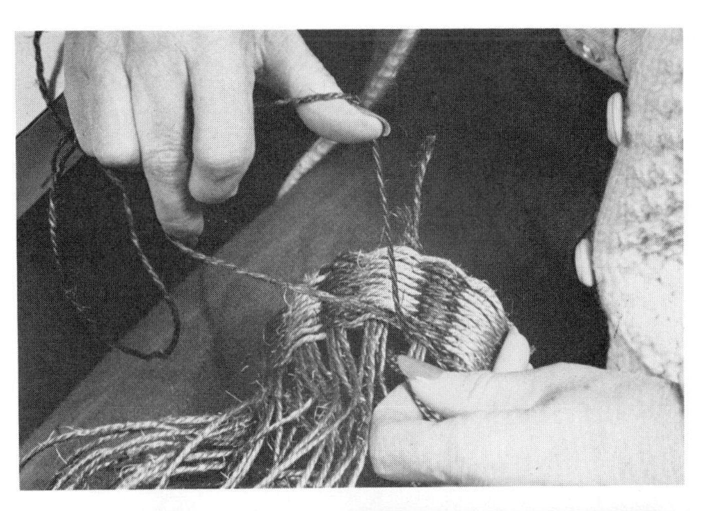

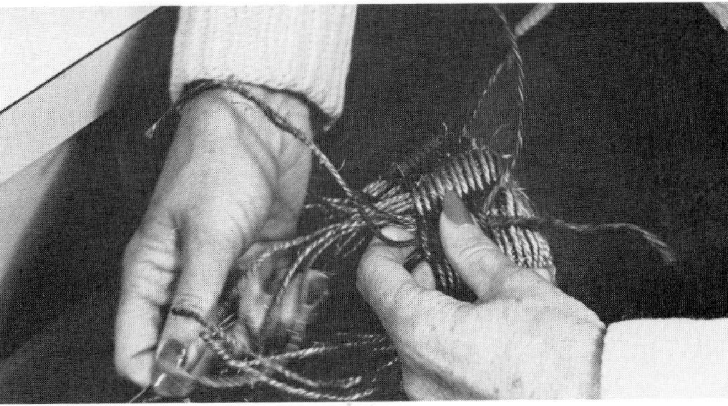

TWINING

Right, above: Spokes are formed of several bundles of jute. The basket is started by using double sets of weavers as demonstrated here. They weave in and out, crossing one another as they twist and alternate between spokes.

Additional spokes are created by sewing lengths of spokes through the twined areas.

"To the East Where the Sun Rises," twined of Greek wool over cord (9" high X 14" diam.). By Louise Robbins. *Courtesy Louise Robbins*

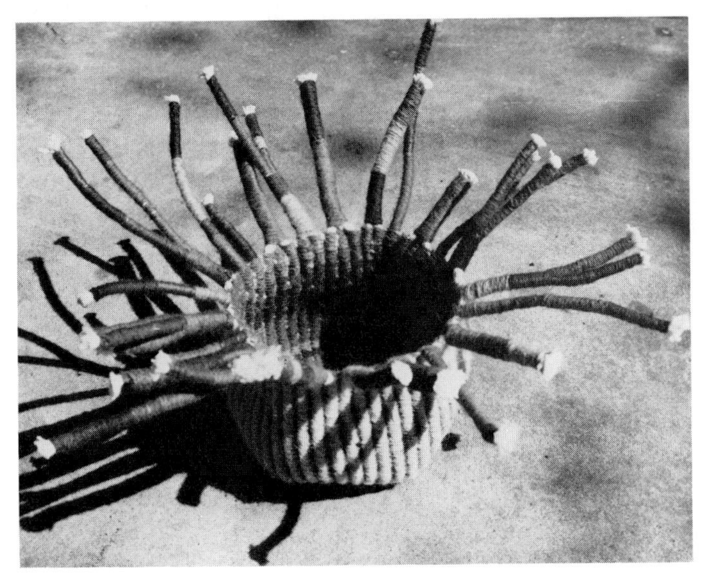

"Fire Spirit," twined of Greek and Irish wool over jute (8" wide X 9" diam.). By Louise Robbins. *Courtesy Louise Robbins*

Twined basket from northern Ghana, Africa.

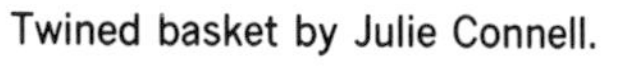

Twined basket by Julie Connell.

Dracena palm leaf twined over peeled eucalyptus tree bark by Lucy Traber.

Lace pitcher-twined basket of palm leaf. By Ed Rossbach. *Courtesy Ed Rossbach*

4 SOFT CONTAINERS

Background

The soft container is not exactly a new concept, although some contemporary treatments of soft materials are definitely innovative. Soft shopping bags, saddlebags, luggage made of heavy and soft woven materials and yarns have been with us for centuries; so have soft stuffed lingerie and handkerchief cases and jewelry boxes. But Renie Breskin Adams's soft teapot crocheted to duplicate every detail of its ceramic counterpart and Joy Nagy's "Bagel" and "Hero" sandwich soft satin purses are definitely more sculptural than functional. Norma Minkowitz's "Horse Heaven Box" is both. The piece can function adequately as a box, but it is much more a work of art. Kay Sekimachi's woven box probably exemplifies the highest levels weaving can reach. In her three-dimensional construction, the body of the box is woven on a harness loom in a quadruple continuous weave, with the lid and bottom woven in a double-weave pickup pattern. The entire form comes off the loom ready to fold into a twelve-paneled, six-sided box with not a stitch to be taken, except where the bottom flaps meet.

Some innovations utilize new materials in an old way, such as stuffing polyester batting into quilted forms. Some take advantage of what new materials, such as vinyls, will do. Other pieces achieve success as art/craft forms through use of conventional materials and processes and atypical subject matter. Carole Austin's quilted cotton Mason jar duplicates the original glass version in every detail, utilizing the traditional quilting process and cotton fabric. Sas Colby creates autobiographical episodes in soft stuffed images contained by a soft stuffed box. The sides of the box are three-dimensional illustrations of her story.

Renie Breskin Adams's "Teapot" is single crochet of cotton yarn. No armatures and no starch are used. Stiffness is achieved by the density and tightness of the stitch. *Courtesy Renie Breskin Adams*

"The Hero," a stuffed satin purse by Joy Nagy (10" X 4½" X 2" deep). *Courtesy American Crafts Council; photo by Bob Hanson*

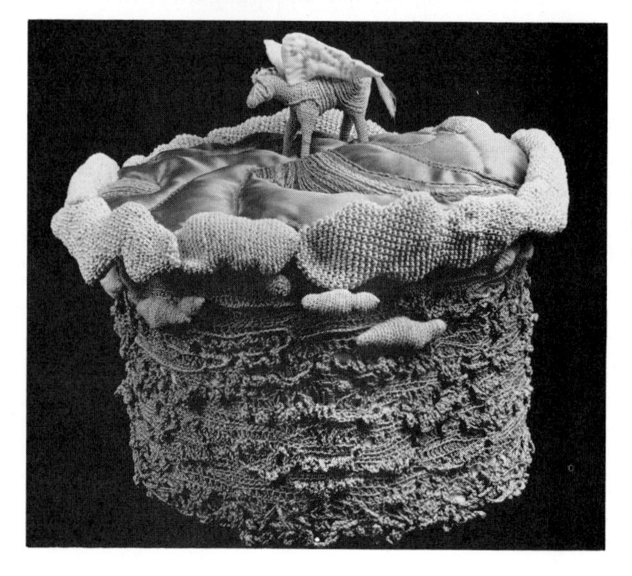

"Horse Heaven," box by Norma Minkowitz in crochet, quilting, and knitting (13" X 16"). *Courtesy Norma Minkowitz; photo by Bob Hanson*

WEAVING A BOX by Yvonne Porcella

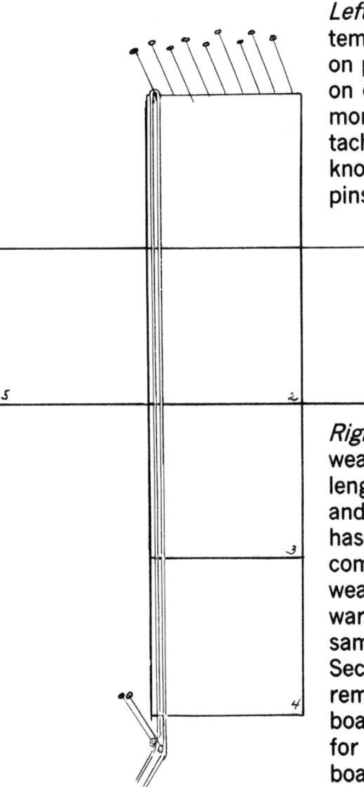

Left: Yvonne Porcella has worked out a simple system for weaving a box. Six equal squares are drawn on paper, as shown here. Place the paper pattern on corkboard or Celotex. Cut warp threads a bit more than twice the length of the pattern and attach them to pins with a loop around one and tie a knot around its opposite counterpart. Press the pins into the corkboard.

Right: After the top square has been warped, weave the first (#1) square. Mount an 18-inch length of yarn on a tapestry needle and weave over and under each warp thread until the first square has been completed. Then beat the weft with a comb to close up spaces and make for a tight weave. When beginning the second square, precut warp thread for the side squares are added with the same spacing maintained as in the first warping. Secure these ends as well. Continue to weave the remainder squares (#3 and #4). Then turn the board around 90° and weave square #5. Repeat for square #6. Remove the weaving from the board by weaving the tied ends back into the weaving.

Below: Sew up the sides of the box and embellish it, if desired. This box is by Yvonne Porcella.

Diagrams and photos courtesy Yvonne Porcella

Right: The woven box when opened contains a stocking doll by Yvonne Porcella. *Courtesy Yvonne Porcella*

KAY SEKIMACHI WEAVES A BOX

The lid and bottom of Kay Sekimachi's box are woven in double-weave pickup in linen, 10/2 for the warp and 10/1 and 8/1 for the weft. The body of the box is woven in quadruple continuous weft weave.

The top and bottom are flaps. The lining is woven continuously along with the outside so there are no seams to the twelve sides of the box.

For stiffening, Stitch Witchery is ironed on with the addition of buckram. The only stitching after the box is assembled is where the bottom flap meets the side.

Kay Sekimachi's box is 6" X 6" X 6".

A vinyl appliqué and trapunto bag by Jeanette Feldman. *Courtesy Jeanette Feldman*

A detail of the bag's surface shows the raised effect in trapunto when appliqués are individually stuffed. *Courtesy Jeanette Feldman*

Carole Austin's "Mason-Dixon" soft jar in sheer cotton batiste stuffed with natural cotton. The "jar" is 4½" and made to the scale of the glass variety. Lettering is machine stitched. *Courtesy Carole Austin*

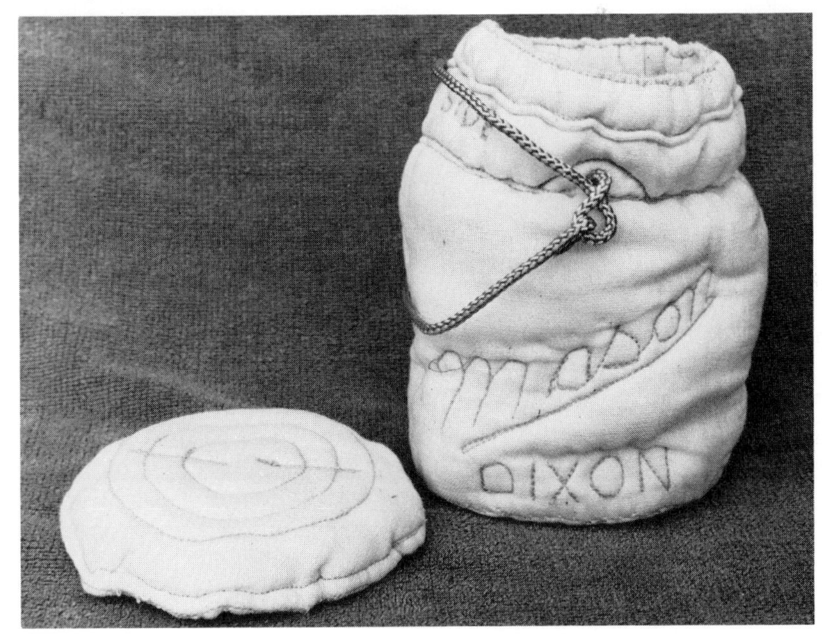

Materials and Construction Concepts

Approaches to creating soft containers vary considerably. Some processes and materials are more frequently used than others. Quilted, crocheted, and knitted pieces are the most popular, perhaps because construction is more direct. One can control the form as it grows three-dimensionally, whereas woven pieces necessarily are frequently woven in flat form and then are manipulated into three-dimensional shapes. It takes a great deal more planning before a weaving can be constructed into a container such as Kay Sekimachi's box. Materials are greatly varied with yarns of all kinds and textures and a plethora of fabrics that can readily be stuffed.

Autobiographical episodes are embroidered on Sas Colby's soft box.

The contents are Sas Colby's family with photographic images for their faces.

Stuffed and Quilted Forms

Stuffed and quilted forms require some kind of flat fabric, such as cotton, thread, and stuffing. Quilted and quilted and stuffed patterns, such as trapunto, do not necessarily depend upon color or dark and light values of the fabric, but rather upon the pattern in relief created by compression of linear areas as stitches are sewn through at least two layers of fabric and one layer of stuffing. Stitches function both to keep the stuffing from shifting around by demarcating areas to contain the stuffing, and, in quilting, to decorate the surface with a slight relief pattern. In trapunto, selected areas are stuffed and the relief is often more pronounced than in quilting.

SAS COLBY MAKES A BASIC STUFFED BOX

To construct a soft fabric box, individual envelopes of fabric are made for each side. Cardboard is used as a template.

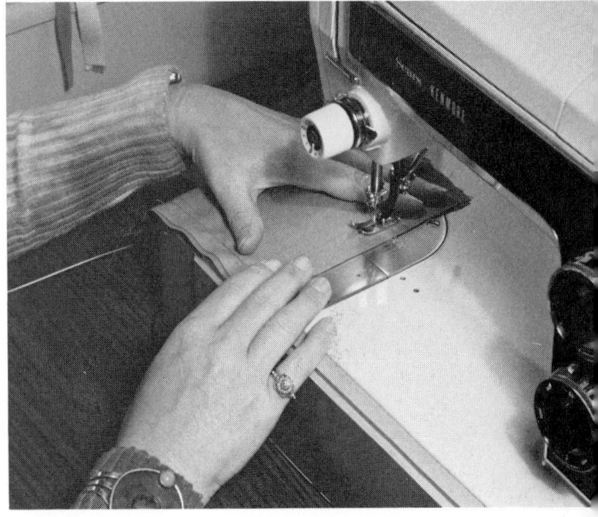

The seams of the fabric envelopes are stitched by machine on three sides, ¼" out from the traced line. Six of these envelopes are needed. The stitching is rounded at the corners.

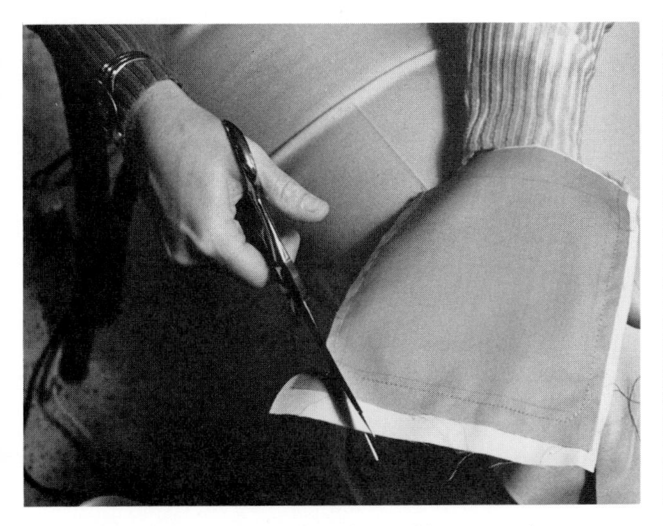

Rounded corners are clipped to enable neat turning inside out without bunching of the fabric. And seams are clipped to about ¼" away from the stitching.

Sas Colby turns the envelope inside out, making certain that the corners are fully turned.

Next, a pattern of fabric mosaic pieces is arranged . . .

. . . and the pieces are joined (via sewing machine). This will become the top of the lid.

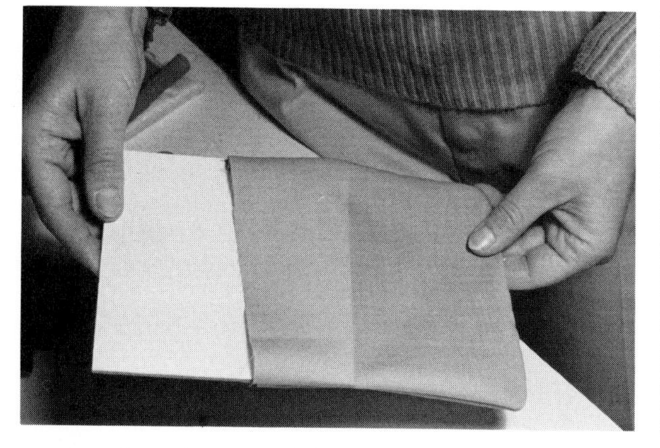

A piece of cardboard, the one originally used as a template, is slipped into the envelope.

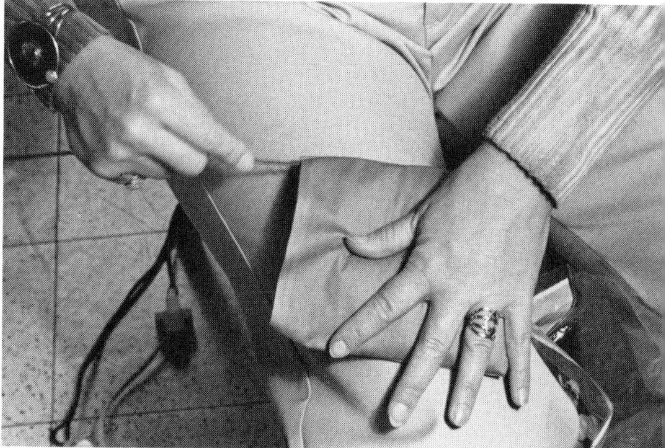

A tool, such as a knitting needle, is used to carefully poke stuffing into the corners and throughout the envelope. Both sides of the sides are stuffed, but only the parts of the bottom and top are stuffed. This is optional; the position and amount of stuffing determine the effect.

All parts, ready to be assembled.

The sides and bottom are attached using straight pins and by stitching along the bottom edge first. Then the side seams are sewn by weaving in and out between the parts with a sturdy needle and thread.

After stuffing the lid, the open seam is pinned closed, sewn, and attached to the box. The box cover is sewn at the back edge—and reinforced by sewing the "hinge" twice.

Sas Colby's completed box.

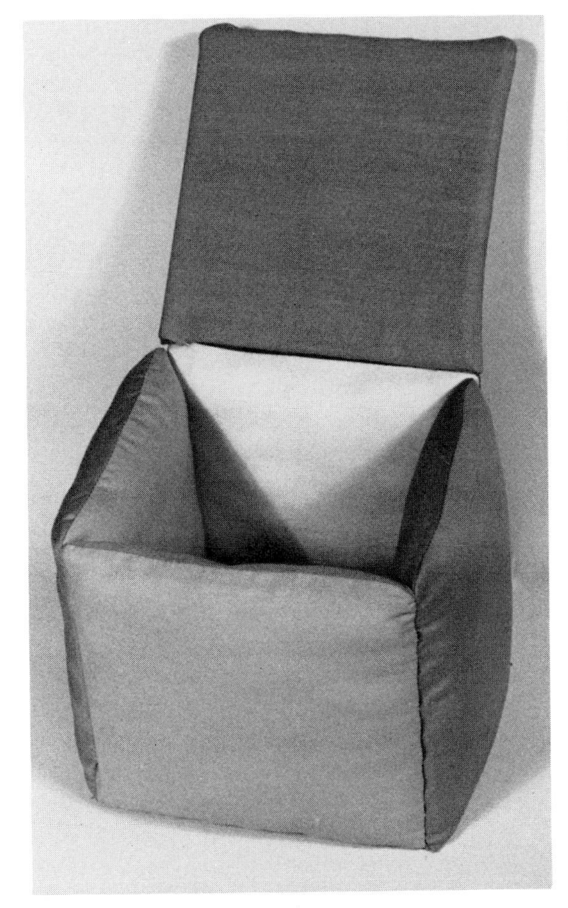

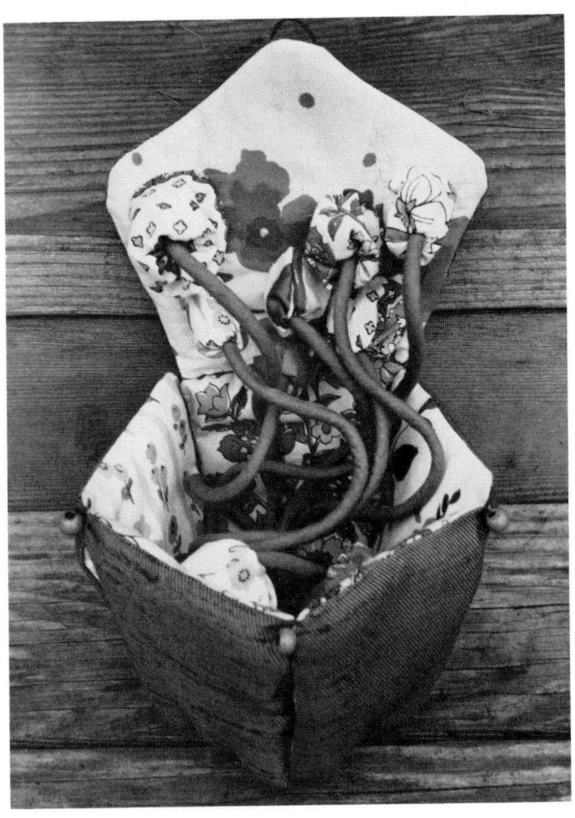

As this fabric pentagon box opens, interior "flowers" grow. Sides of this pentagon unbutton and open outward. By Marcia Morse. *Courtesy Marcia Morse*

Fabric Choices

Early quilt makers had very little choice of textiles. Every innovation in the textile industry brought changes of approaches and images. Today there is an embarrassment of materials, colors, textures, and patterns. Making a choice amidst the many available to us can be very difficult. Whereas once delicate-looking fabrics such as laces and chiffon were also fragile, today fragile-looking fabrics can wear well in heavy-duty applications. The advent of synthetics has revolutionized the textile industry. Mind-setting conventions of the past are no longer relevant. We have plastic fabrics such as Ultrasuede, the polyurethane/polyesters, and a wide range of double-knit fabrics. Heat processes now replace stitches in quilting, and in some cases thermoplastic fabrics are molded into permanent forms.

Stuffing Choices

The choice of stuffing material has expanded too. Fillers were once limited to cotton or wool batting, down, and feathers, kapok, straw, rice, and scrap fabrics. Today, besides the traditional fillers and the use of scrap, there are many densities in flexible and rigid foams, polystyrene pellets, and batting, polyester (Dacron) batting, polyethylene bubble "paper," cork, rock wool insulation, terrycloth, felt, and so on. Even use of scrap materials can be extended to include nylon stockings, and the varied types of stuffing from old mattresses and pillows to discarded recycled materials.

Putting It All Together

Quilting is a very old technique that uses stitches to attach and hold a central layer of filler (batting, wadding, stuffing) between two layers of fabric. Stitching throws the fabric into relief as it pinches the layers of fabric together, transforming the surface into a play of light and shadow. The pattern formed by stitches can be immensely varied, from overall geometric and calligraphic lines to meandering, free-form designs.

Variations in type and amount of stuffing also create great diversity. A stuffed soft form is three-dimensional and often squeezable because of its puffed-out appearance. Quilting, trapunto, cord-quilting, and stump work—all are stuffed works in varying degrees of relief.

Trapunto is high relief worked through densely packed pockets formed by layers of cloth. Varied effects are achieved through the use of different kinds of cloth, too. The Italians use sheer fabrics, the French, heavier materials.

Generally in trapunto, the bottom layer of fabric is a loosely woven lining. By pushing and drawing aside the warp and weft so that a temporary hole is formed the stuffing can be inserted into areas between the two layers with a blunt tapestry needle or crochet hook. After the proper amount of stuffing is metered into the area, the threads of the lining are pushed back to normal position, covering the stuffing. Where the liner threads cannot be manipulated, a slit is made in the lining for entry of the stuffing. This opening is later sewn closed. In some instances, a patch is made on the underside to contain the stuffing.

Individual units are sewn, stuffed, and assembled into a cylindrical form by Lida Gordon. *Courtesy Lida Gordon*

Close-up of the opening of Lida Gordon's container. *Courtesy Lida Gordon*

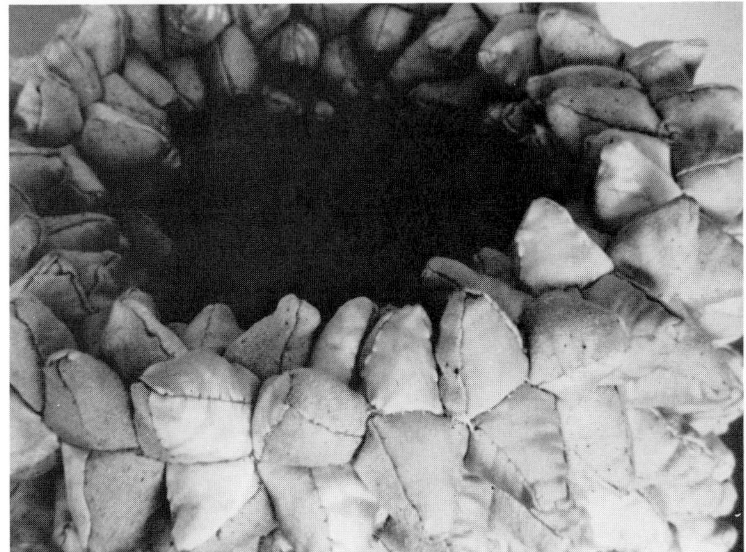

Maud Guilfoyle draws directly on smooth surfaced unsized cotton (percale preferably) using a crow quill point pen and Pelikan drawing ink (Castell TG or India ink also work). She is careful not to pause but to draw with a continuous line. Pauses cause the ink to blot as the fabric absorbs excessive amounts of ink.

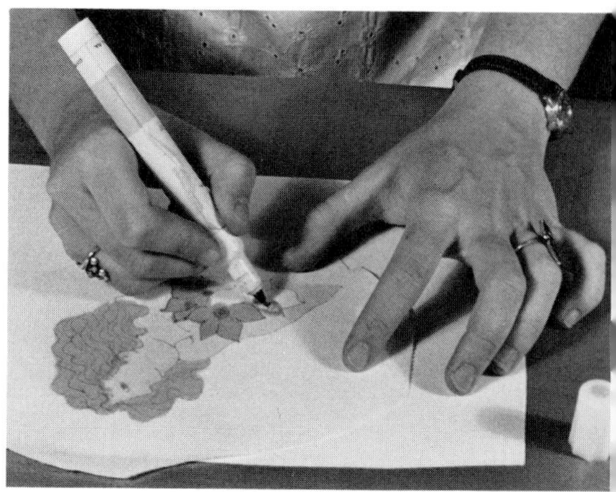

She uses AD markers (Cooper Color, Inc., Jacksonville, Florida) for coloring. These are more colorfast and do not bleed excessively. To fix color, she soaks her work in vinegar and water for ten minutes, air dries to dampness, and steam irons the piece for five minutes. (Some markers dissolve the Pelikan ink outlines.) When coloring with AD (or Sharpie, Sanford Co., markers), Maud Guilfoyle does not go to the line but just below it to accommodate for some natural spread of color to the line.

Maud Guilfoyle's soft box with drawings.

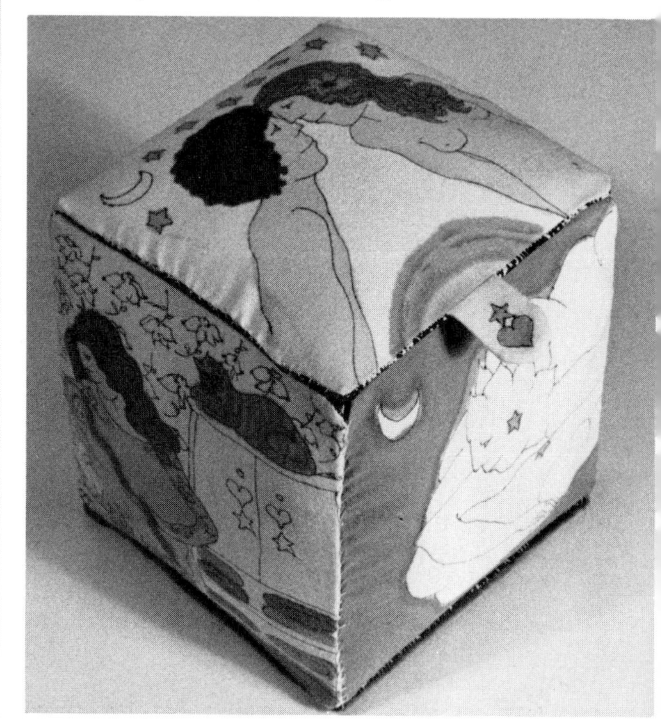

Three layers are assembled for quilting the drawing part of a box lid; Poly-fil (Fairfield Processing Corp., Danbury, Connecticut 06810) is sandwiched between a lining and the drawing.

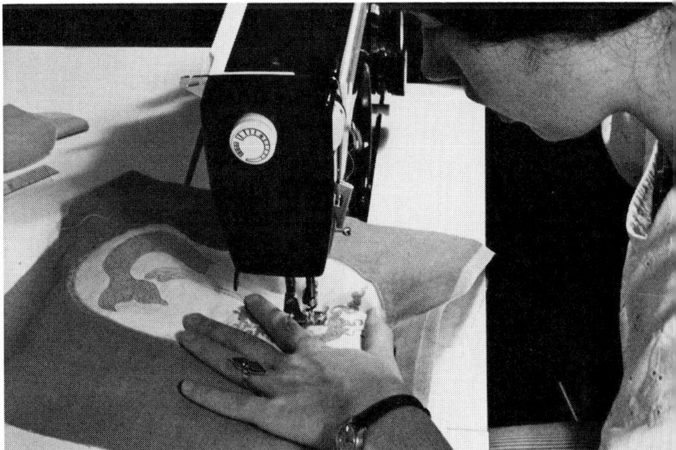

Maud Guilfoyle stitches the contours of her drawing on a sewing machine. This creates a quilted effect that further defines the clouds.

The fabric frame around her drawing is further stuffed and then attached to a stuffed and lined box.

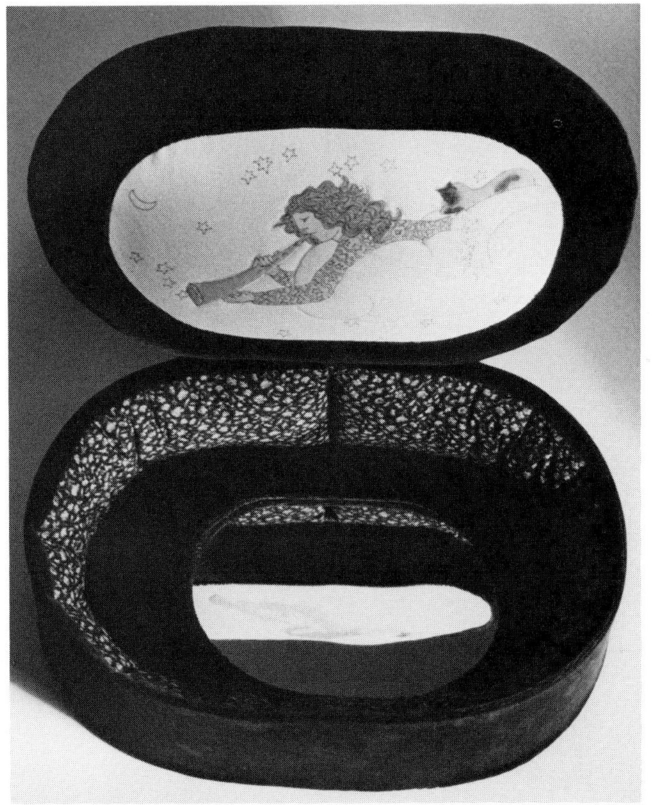

The box interior contains another drawing and at the base is a mirror.

Above: "Angel" bag by Elizabeth Gurrier is quilted and stuffed. *Courtesy Elizabeth Gurrier*

Top, right: Anne Swan's "Marsupial Group," stuffed, painted containers with pouches, hanging from a rod. *Courtesy Anne Swan*

Center, right: Two bags by Joy Stocksdale: machine-quilted satin on the left and hand-quilted and embroidered on the right.

Bottom, right: Stuffed satin "Bagel" bag by Joy Nagy (4½" diameter, 1½" deep). *Courtesy American Crafts Council; photo by Bob Hanson*

Below: A loose white velveteen trapunto-stitched drawstring soft trunk by Anne Swan. *Courtesy Anne Swan*

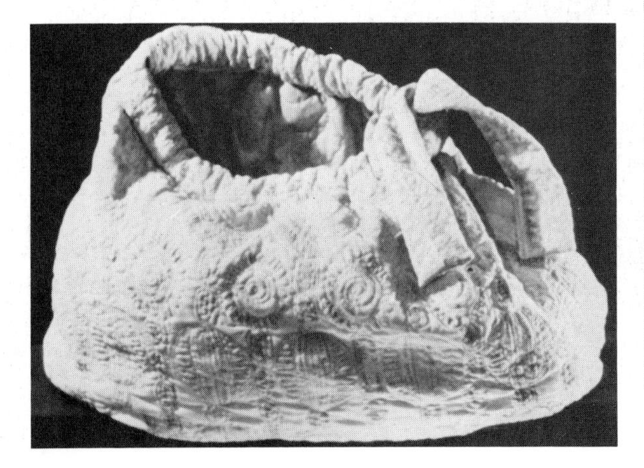

A stuffed turtle bag opens up to reveal a turtle toy with jingle-bell eyes and a scarf. By Lisa Drumm. *Courtesy Lisa Drumm*

Outline of the Procedure for Making a Stuffed Form

1. Plan your design in detail
2. Cut patterns for the form
3. Trace or cut pieces around pattern for each shape
4. Assemble cut forms into a sandwich, using a filler if it is for quilting; or without the filler if it is for trapunto or to be a 3-D stuffed piece
5. Sew elements together:
   a. For simple stuffed 3-D form:
      1. Sew around unit, leaving opening for turning fabric inside out
      2. Stuff unit
      3. Sew opening closed
   b. For complex stuffed 3-D form:
      1. Sew around each unit, leaving opening for turning fabric inside out
      2. Stuff unit
      3. Sew opening closed

    4. Repeat for each unit
    5. Attach units together
c. For quilted form:
    1. Make a sandwich of two layers of fabric with filler in between
    2. Draw quilting pattern on top of bottom piece
    3. If possible, attach to a quilting frame or portable hoop
    4. Proceed to quilt in a plain running stitch, up and down through all layers, or sew on sewing machine
    5. Manipulate the piece and attach edges together or hem edges closed
d. For trapunto:
    1. Draw design on one of the layers of fabric
    2. Put both layers together
    3. Sew around outline with a tight running stitch, or sew on sewing machine
    4. From underside (lining), force an opening
    5. Stuff filler into area, compacting it as much as necessary
    6. Close opening
    7. If needed, shape form by joining the parts

Yarn and Fiber Forms— Knitted and Crocheted

Materials used for fiber containers include a vast assortment of yarns and cords. Rayon, cotton, linen, wool, Dacron, nylon, jute, hemp, and sisal are just a few. Besides using yarn, roving, fur strips, plastic cording, strips of metal, wood, vinyl, and felt are among the usable materials.

Some pieces require stiffening. This can be accomplished by stuffing the parts or by using cardboard or interfacing materials such as Sta-Shape, Milium, Pellon, buckram, and so on. Sometimes, armatures of aluminum wire or clothes hangers are necessary. One way to construct an armature is to cut several lengths of cord (or use a heavy cord) that have been wrapped or wound tightly with another yarn, as in coiling. This makes the core cord stiff and provides for a decorative armature, particularly if the wrapped color is coordinated and permitted to show, as in a basket.

The covering for a three-dimensional knitted or crocheted container is built up by increasing or decreasing stitches as shown in the process diagrams.

Stuffing or armatures are most often placed into these forms later because it becomes very difficult to crochet or knit when an armature or stiffening is in place.

Knitting and crocheting are needleworking processes because each requires use of one or more needles. The size of needle for each depends upon the thickness of the yarn or fiber. Multiple needles or a circular needle can be used for knitting, particularly where tubular forms are to be made. Once the process of increasing or decreasing within the loop structure has been mastered, any kind of shape can be made. Parts can be crocheted or knitted and then assembled by sewing them together.

Knitting is a very old craft that dates back centuries, but very little is known about its origin. Flat knitting is worked on two needles throughout. Circular knitting, as indicated earlier, is worked on a circular needle or on sets of four or five needles, each having points on each end.

Stitches vary with how knit and purl stitches are worked. Patterns are created by interchanging of knit and purl stitches and how the fiber or yarn is wrapped on the needle.

Increasing stitches is accomplished by knitting and purling in the same stitch, thus making two stitches out of one; this can be increased to three or four by repeating the process in the same stitch, picking up the loop that lies between the stitch just worked; casting on another stitch at the end of a row; or double increase at the end of the row as indicated in the first process.

Ways of decreasing stitches in knitted fabrics are by knitting two stitches together; slipping one stitch onto another by knitting a stitch and passing the loop over the knitted stitch; or combining both techniques of knitting together and slipping stitches one over the other; or knitting three stitches together.

To cast off, always maintain the same tension.

One can create round or other types of shapes from a flat knitted piece by sewing parts together or by combining with crochet.

Crochet

Crochet work is not so old a craft as knitting. Only one needle (with a hook) is used in crochet and the mesh is firmer and does not stretch as much as knitting (but it consumes more yarn because of its firmer structure).

Simple crochet is based upon a chain. By joining the chain we get a loop. To increase, more chains are added, or double or triple (or more) crochet into the same loop, or cast more yarn around the needle. To decrease, skip a stitch or more, or slip one stitch into the other.

Sometimes the piece is turned around and worked on both sides, other times stitches are made only in one direction, ending the thread at the end of each row. Patterns are created that way or by skipping stitches, crocheting chains within stitches, doubling stitches, going through both threads (d.c.)* or into just the top or single (s.c.).* Loops can be created by crocheting around a stick and then crochet a line or two at the top of the loops after the stick has been removed; thread can be pulled through several stitches at one time; or different colors of thread can be introduced into the base color.

*d.c., double crochet; s.c., single crochet.

Left: A man knits his own bag. Lake Atitlan area, Guatemala, Central America.

Above: Standard knitting procedures are employed.

Right: A typical knitted woolen bag from that area of Guatemala.

Wool and natural color dyes are used for this Arauchan bag from Colombia, S.A.

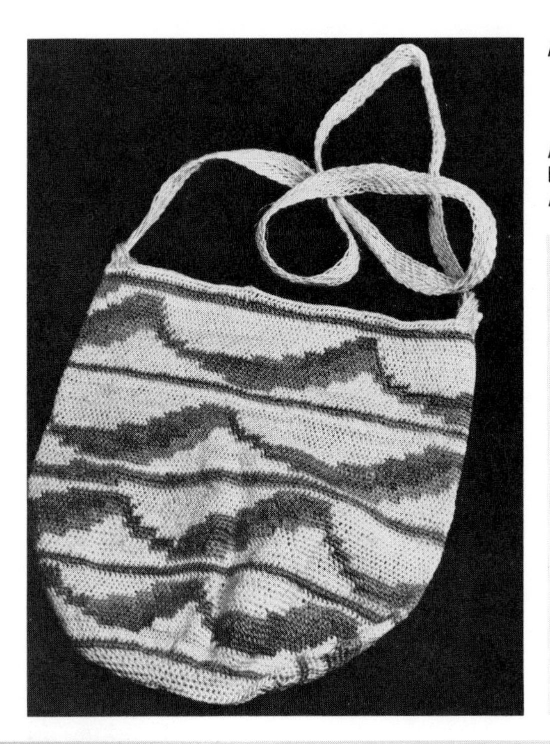

Another version of an Arauchan bag in a flaxlike yarn.

An open crochet pot in shades of rust/purple, lined in burgundy with golden pheasant feathers. *Courtesy Nancy M. Piatkowski; collection C. Jackson Brockette, Jr.*

The beginning of a crocheted pot by Nancy M. Piatkowski. *Courtesy Nancy M. Piatkowski*

"Speckled Enamel Coffeepot," crochet by Renie Breskin Adams. *Courtesy Renie Breskin Adams*

Opposite page:
Top: A crochet miniature, "Still Life with Picture on the Wall." Note the scale as indicated by the pen. By Renie Breskin Adams. *Courtesy Renie Breskin Adams*

Center, left: Teapots in cotton by Renie Breskin Adams. The relative lack of stretch in cotton yarns and the geometry of regular basic stitches contribute to the sturdiness of these containers. *Courtesy Renie Breskin Adams*

Center, right: "Bird Seed Bag" (7" X 17"), crochet by Norma Minkowitz. *Courtesy Norma Minkowitz; photo by Bob Hanson*

Bottom, left: "Victoria's Handbag," a gargantuan woolen handbag, crocheted by Thomas Charles Siefke (45" X 36" X 36"). *Courtesy Thomas Charles Siefke*

Below: "Bridal Box" (18" X 14"), crochet and beads by Norma Minkowitz. *Courtesy Norma Minkowitz; photo by Bob Hanson*

Crochet container of wool, stuffed, quilted, and supported with plastic stays to keep it upright. By Diana Schmidt Willner. *Courtesy Diana Schmidt Willner*

Procedure for Creating Containers by Crocheting or Knitting

1. Plan and design piece
2. Make patterns for contours, particularly if working with flat areas at first, or establish measurements
3. Decide on stitches for texture and pattern
4. Proceed to knit or crochet
5. If stiffener or armature is needed, form that
6. Form container by joining parts and, if necessary, insert armature or stiffener (before or after joining parts as the design may require)

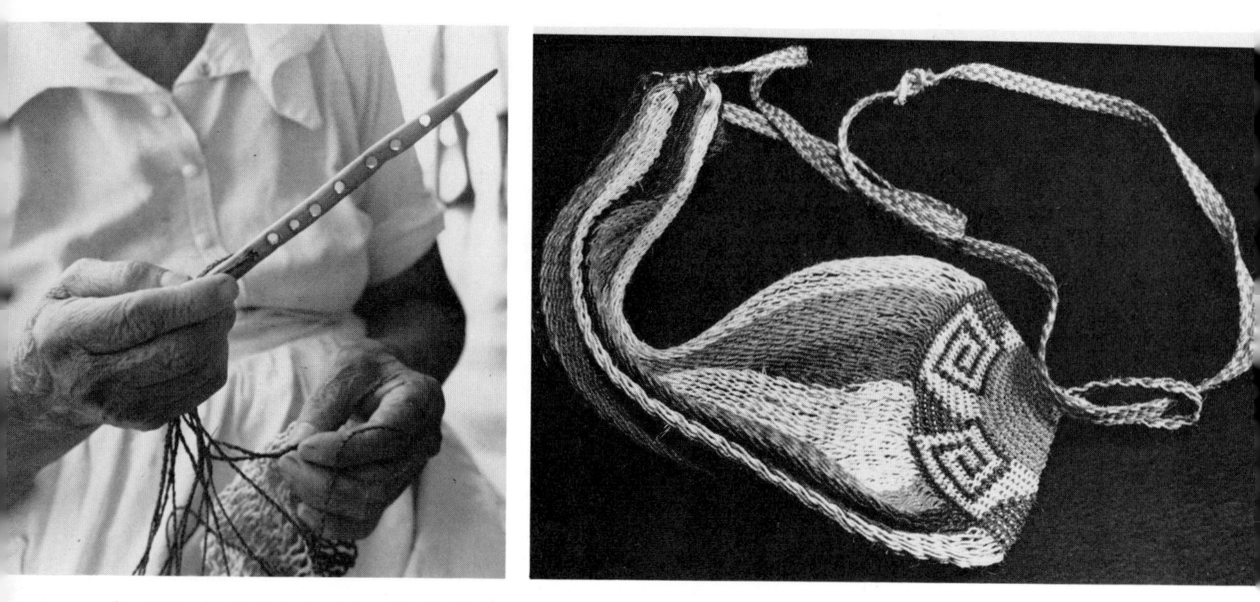

A netting needle is looped in and out of previous loops.

A netted basket from Oaxaca, Mexico.

Woven Containers

Woven containers are not conventional forms but rather a departure from the usual weaving of a cloth. Weaving is an ancient art with origins dating back to prehistoric times. Weaving techniques, types of looms, kinds of yarns, types of patterns, are extremely varied because its history is vast.

Each textile is constructed of two sets of threads that cross each other in various ways but with some kind of warp and weft. The warp stretches lengthwise on or off the loom, and the weft runs in and out of the warp, usually horizontally.

There are a great many variations on the theme of weaving from darning to finger weaving, Indian braiding, two-harness and backstrap loom weaving, card weaving, multiple harness loom weaving, and so on.

Weaving is essentially a two-dimensional art form. In order to create containers, the weaver has to build in the constructs to create a three-dimensional form. It is possible to double weave using multiple harnesses and then fold or stuff the form; or to weave strips that are then manipulated in some way into a three-dimensional form.

One can also construct a loom on rings (by stringing the warp on rings) or around large tubes (notching the tube top and bottom) and weaving *around* a warp rather than back and forth.

Once released from the bind of weaving only flat surfaces, license can be taken with the process, even to combining weaving with crochet, macramé, and so on. Certainly, once a flat weaving comes off the loom it can be curved, folded, cut, sewn, and gathered into a variety of container forms.

The process of weaving is too vast to be dealt with here. If you are not familiar with weaving processes, it is best to refer to one of the references noted in the bibliography for in-depth instruction.

Procedure for Creating Woven Containers

1. Plan and design your form. Decide on size, texture, pattern, and color. Cut any needed patterns
2. Set up loom
3. Weave
4. Remove fabric from loom and construct container form

A woven saddlebag by Julie Connell.

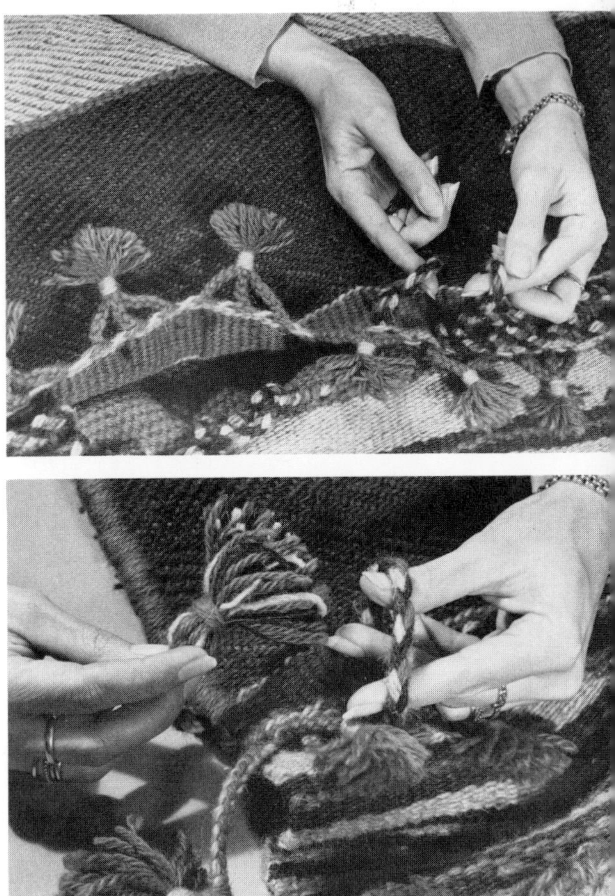

Right: To close the saddlebag, tassels pass into loops all along the opening, much like a chain stitch. . . .

. . . until the final tassel hangs on the side.

Knotless netted pot with stocking doll heads on lid and inside. By Yvonne Porcella. *Courtesy Yvonne Porcella*

A small saddlebag of henequen from El Salvador, Central America.

Strips of cotton belting woven on a narrow backstrap loom in Santo Tomas Jalieza, Oaxaca, Mexico, are sewn together into a bag. The symbols are animistic.

Other "Soft" Containers

Embroidered and appliquéd containers also have a long history; ladies' purses, shopping bags, lingerie, and handkerchief cases are all part of the traditions of embroidery and appliqué. Stump work sits between trapunto, inasmuch as it is stuffed or three-dimensional, and embroidery itself, because the stuffing (cording) is covered with embroidery.

Knotted and hooked processes usually reserved for rugmaking have been processed into container forms. The "carpetbagger" of post-Civil War days carried a piece of luggage made of carpet. Certainly this type of fabric would withstand heavy wear. Since the texture usually is coarse and heavy, these hooked and knotted rug techniques are usually reserved for large container forms.

STUMP WORK BOXES by Joy Stocksdale

Stump work (raised stitches) is done on 12-mesh interlock needlepoint canvas. Dacron upholstery cording (welting) is used as a filler and is loosely tacked into a design on the mesh.

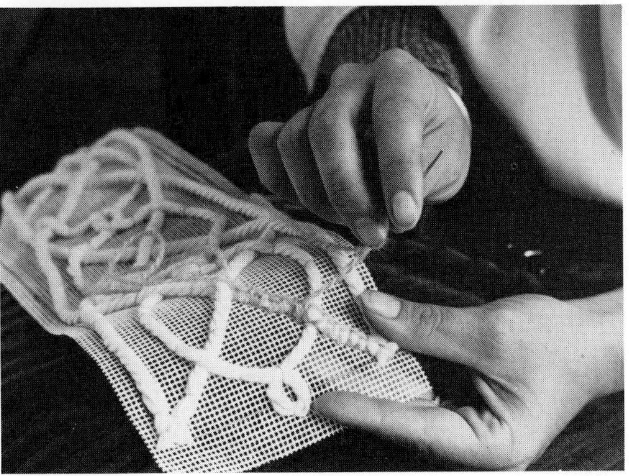

The end of a two-ply tapestry wool is stitched around the cording in ¼" intervals. The stitch acts as the warp.

Top: Then a weft is woven in and out along the cording until the entire cord is covered and yet retains its round shape. As each weft thread is completed, it is tacked at the bottom so that it covers the end of the cording. The rest of the spaces between the cords (the background) are filled with needlepoint.

Center: A binding stitch is whipped around the edge to a lining. The binding stitch is a cross-stitch that weaves into the next cross-stitch and through all the layers of fabric.

Bottom, left: A supporting box is slipped into the covering. Holes were punched previously around the base of the box in order to be able to sew the box to the covering.

Bottom, right: Dacron stuffing is used in the interior—for sides and bottom the lining covers it.

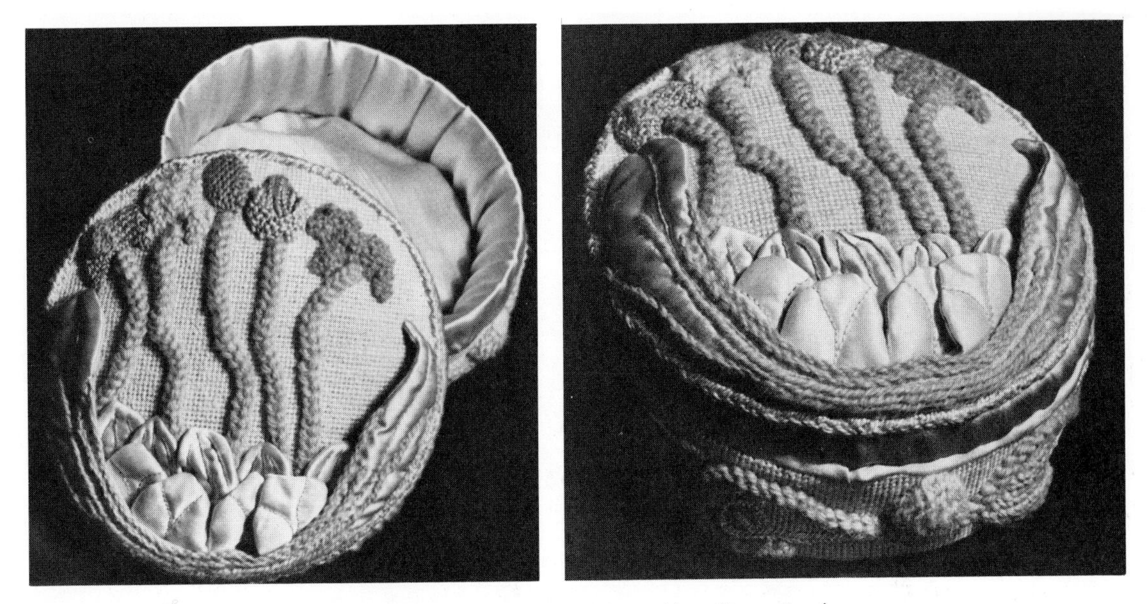

Joy Stocksdale's stump work box with satin appliqué.

Appliqué over a soft, stuffed velvet box by Carole Austin.

Appliqué and embroidered bags by the hill tribe people of northern Thailand.

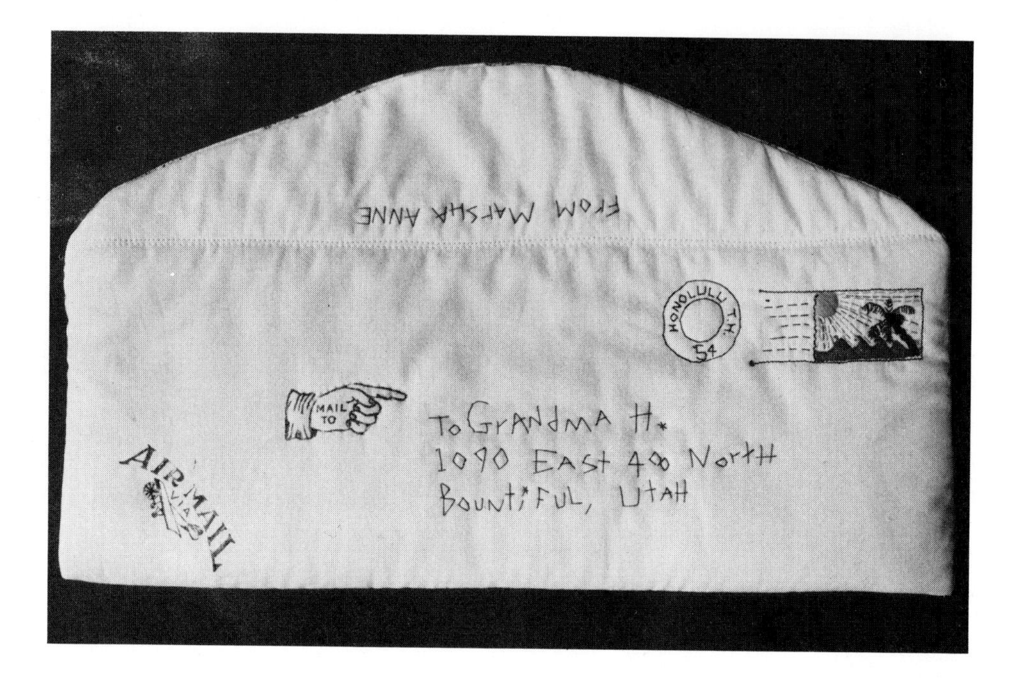

Marsha Anne Isoshima's envelope—stuffed and embroidered. *Courtesy Marsha Anne Isoshima; photo by Sally Mead*

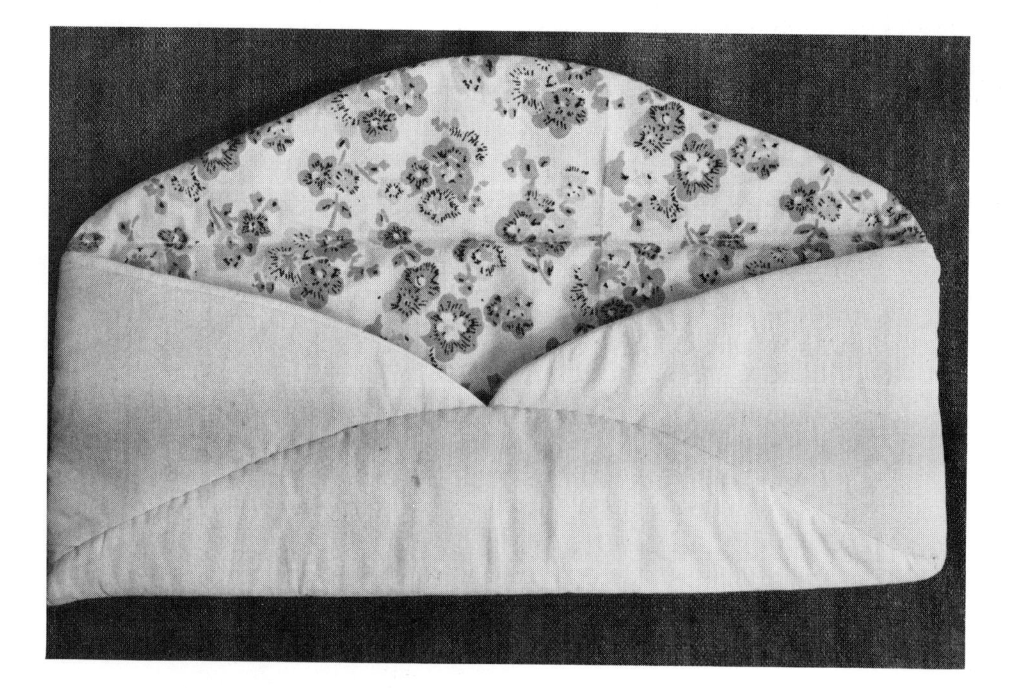

LEATHER CONTAINERS 5

Leather is an incredible material. As a skin, it is the outer covering or bone-container for all the animal world. As man-made clothing, it conforms to the body like a second skin.

Leather can be supple, stiff, scaly, smooth, solid, strong, delicate, heavy, light, thick, and thin. Some leathers have lasted for centuries, others have crumbled in a few years. Leather can be the structure itself or just a covering. It can be sewed like fabric; cut, scored, and glued like paper; carved, laminated, and nailed like wood; hammered and tooled like metal; painted like canvas; woven and tied like yarn; formed like papier-mâché; molded like felt; and dyed like textiles.

Basic Types of Leather

The history of leather can be traced through both its products and the complex tanning processes. Tanning of hides and skins arrests decomposition, increases their strength, makes them more pliable, and keeps leather from becoming soluble in water, and from drying out in air. Oak tanning dates back to the Bronze Age, sometime around 2500 B.C. Not only is that process still used, it has not changed significantly in all this time. In fact, only very slight changes in the tanning process have been made. But in the nineteenth century, a major development occurred: the discovery of chrome tanning. Rather than using tannin, a vegetable tanning agent often found in tree bark, chromium salts, a mineral, was used. The original results were stiff, but this process took a few days rather than the months necessary in vegetable tanning, and permitted the use of various colored dyes. Later on, with the introduction of soap and fat into the tanning process, chrome-tanned leather became water resistant, durable, and flexible, and a likely competitor to vegetable-tanned material.

An Italian leather-covered box with incised and stamped design (8 1/16" diam. X 1⅛" high) that dates from the fifteenth century. *Courtesy The Metropolitan Museum of Art, Rogers Fund, 1950*

Proper Leather for Particular Uses

Despite the excellent chrome-tanned leathers we use today, oak-tanned (and other vegetable-tanned) leathers are still very much with us. These are the best leathers for forming and carving. Unlike oak-tanning, chrome-tanning is an open tannage process creating a more porous material; therefore, chrome-tanned leather cannot be used for forming. When wet, the leather becomes stiff rather than flexible and will not hold its shape.

To determine whether a leather is chrome-tanned or vegetable-tanned, look at its cross section. A bluish gray streak in the center layer indicates chrome-tanning. Vegetable-tanned leather is more uniform in color.

Another test is to moisten the leather with warm water and then fold it. The vegetable-tanned piece will feel slippery but will hold a crease. Chrome-tanned pieces, on the other hand, will feel drier and spring back.

It is most important to know which skins to use for particular applications, even though specific properties of leather can be controlled to a certain extent by the type of tannage. The intrinsic structure of each leather imposes certain limitations. Goatskin cannot become calfskin, nor can they always be used interchangeably.

The following list should serve as an indicator of which leathers work best for specific products. It is presented here with some reservations because there are always exceptions to common practice that turn out particularly well.

Boxes, bowls: 10 ounces and up, vegetable tannage; usually cowhide, steerhide. (As coverings over wood or another material): 2- to 3-ounce vegetable tannage.

Briefcases: 4 to 6 ounces; heavy cowhide, latigo, steerhide.

Handbags: 4 to 10 ounces, depending on whether the bag is to be lined or molded; lambskin (garment suede), medium weight split cowhide, and some novelty leathers (e.g., snakeskin), if lined.

Wallets, envelopes, and similar containers: 2 to 4 ounces; pigskin, calfskin, cowhide, goatskin.

Linings: 1 to 5 ounces, depending on what it is to be combined with and its use. Skiver, splits, and suedes do best (linings can minimize stretching).

Sacks (or other soft containers): 2 to 4 ounces; buckskin, lambskin (garment suede), lightweight cowhide, pigskin, sheepskin (Cabretta), goatskin, scraps of various leathers, top grain, and splits.

This Fulani saddlebag from Africa is made of tawed leather, which is a cosmeticlike treatment of leather and not true tanning. Temporary preservatives are employed in tawing. Since no tannin, salt, or alum is absorbed chemically into the hide, the process is reversible.

Chrome-tanned leather is used in making this bag by Walter Dyer. *Courtesy Walter Dyer*

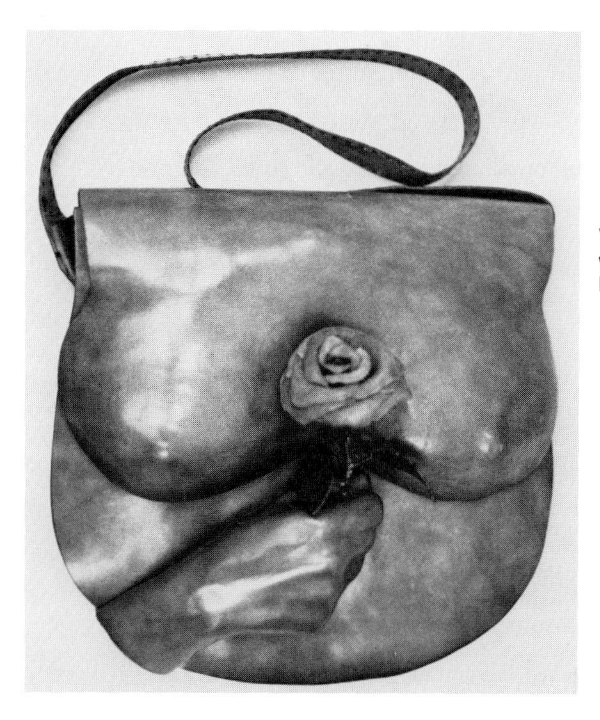

Vegetable-tanned leather is necessary where wet-forming is employed, as in this wet-formed bag by Marcia Lloyd.

Basic Tools

Leather can be worked with minimal tools or, of course, in a workshop where every tool and machine is at hand. The only difference is that the processes are easier and the end product could look the same. For example, one can utilize a tree stump as a cutting table as well as a butcher block.

To begin, some area should be reserved for laying out a hide and cutting the pattern. Another space should be available for hammering and pounding. Some accommodation should be made as well for storage of supplies and the use of dyes and coatings. And a clean, dry place (not hot) should be available for storing leathers (laid flat or individually rolled).

The basic tools are single-edged razor blades, sharp knives, 7½" to 8" sharp shears for cutting, an awl for piercing holes; a ball-point pen for marking lines and patterns; a ruler for measuring; a hammer or mallet for striking, fixing seams, and attaching rivets; a brush; daubers or cotton swabs for applying edge coatings; sponges or soft rags for applying dye, wetting down leather, and polishing; a square for getting the proper angles; a fork or spacing wheel to indicate spacing for holes; a revolving punch or thonging chisels for forming various holes and slits in leather for sewing or thonging.

Of course, the list could be much longer. The basic machine tools are a buffing wheel, an electric sander for carving and roughing surfaces, and a drill press (for drilling holes). For some designs a jigsaw, band saw, or saber saw can save time when cutting thick leather (5 ounces and up).

A Neolite sheet (see your shoemaker) placed under the leather makes a good cutting base to keep from dulling sharp tools and scratching counter tops.

Threads of various kinds, such as heavy-duty carpet thread, linen bookbinder's thread, waxed threads, and cotton-covered Dacron, are excellent for sewing leather parts together. The choice of thread is determined by the function of the container and type of leather being used. Dacron thread can be used to sew garment leather, but is not advisable when using latigo for a briefcase. Thonging, thin strips of leather and plastic in various colors and thicknesses, is used for attaching parts, too.

Leather cement, such as rubber cement, Barge, or other leather-specific adhesives, is also useful for attaching parts, laminating layers, and adhering linings.

Dyes are important for coloring leather, and polishes and waxes protect the top grain.

Although each item will not be necessary for each project, sooner or later all these tools and supplies will be put to use.

COVERING A WOODEN BOX WITH LEATHER

A pattern is made for the covering. It is tested against the wooden box to make certain all dimensions correspond. Then the pattern is placed face side down on the back of the leather, which is cut with a sharp knife. Masking tape or Plasti-Tak is used to keep the pattern from moving.

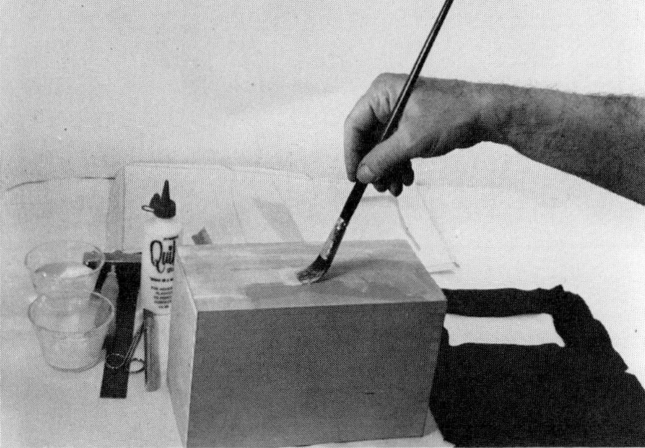

Velverette or a similar vinyl glue, such as Quik, is brushed on the wood and leather. Barge Cement (a specific adhesive for leather) can be used as well.

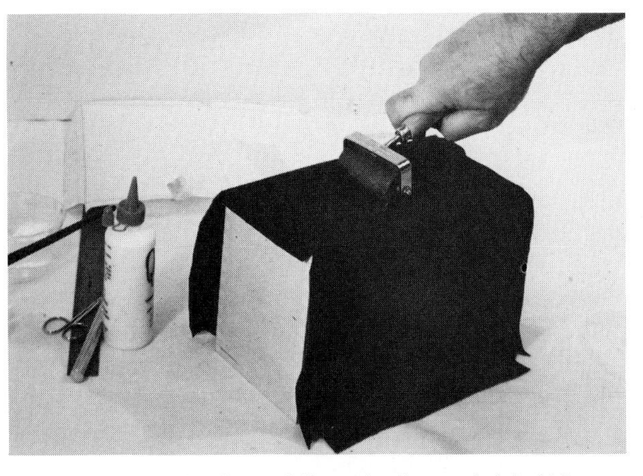

The leather is carefully put in place and air bubbles are rolled out with a brayer.

Areas where the leather has to fit around curves are cut so that there is no bunching or wrinkling. Where edges overlap, excess is skived away so that the two pieces are as thick as one when joined.

With the addition of a lining, a decoupaged panel, and hardware, the purse is complete.

An embossed leather wastepaper basket by Froelich Leather Co. Leather can be used as a covering over almost any material.

This Spanish leatherwork box from the eighteenth century contains a chess set (5½" deep and 14⅝" square). This style of leatherwork originated in North Africa, was brought to Spain by the Moors, and then by the Spanish to Mexico. *Courtesy Metropolitan Museum of Art, Gift of Gustavus A. Pfeiffer, 1948*

Basic Processes

The fundamental operations have come down from times past when tools were few and simple. These few essential processes are elemental to the most complex containers.

Pattern Making

After designing the container, patterns should be cut for each piece. If there are any doubts as to how well parts will fit, artisans usually cut and sew fabric into the intended shape as a trial run. Making adjustments in this way reduces the potential for error.

Paper patterns are translated into leather by tracing around each pattern shape with a ball-point pen or the sharp point of an awl.

Cutting Leather

Sharp knives, single-edged razor blades, or heavy weight shears are excellent for cutting leather. Any cutting tool with a sharp edge will do, but dull blades tear the leather and leave a ragged edge. If the leather is thick and heavy, a handsaw or saber saw may be necessary.

When cutting, craftsmen usually place the leather wrong side up over a heavy piece of leather, hard rubber, vinyl, or Neolite sheet, so that when the cutting edge passes through it will not catch on or be dulled by the counter top. Steel rulers or T squares make good guides for straight edges. And the best cutting technique is to hold the knife at a 45° angle and press down firmly and continually while cutting. Always apply enough pressure to cut through the first time around.

Skiving

If several edges are to meet in a joint, they should be thinned—or skived—so that when overlapped the two together are only slightly thicker than the original. Of course, only the areas to be adhered need be thinned, and skiving is not necessary when using thin leathers.

Skiving is usually done on the flesh side (underside) with a very sharp knife starting about 1/2 inch in from the edge and the blade is held at a shallow angle to make short cuts until the proper amount has been skived away.

MARC GOLDRING'S BOX

The head knife is used almost exclusively by Marc Goldring for cutting and skiving heavy leather.

An edge beveler in action. Marc Goldring trims the edge of heavy cowhide, which will become the side of his box.

Above, left: When parts are glued together, as for the box interior, areas to receive glue are sanded.

Above, right: A "V" groover is used to cut away a sliver at corners where the leather is to be folded to form a box. After assembling, the box is waxed and polished.

Left: One view of Marc Goldring's box.

Below, left: A view of the back. The hinge is hand sewn onto the back with pin made of wood.

Below: The open box showing the interior as a separate insert.

Process photos by Marc Goldring

A decorating process invented through play by Marc Goldring's daughter. Leather glue (rubber cement type—Master's Cement) is dripped onto the surface of clean leather. The leather is dyed and the glue is then removed. Areas protected in this resist process become raised (through swelling); background areas are lower. The longer the glue is left on, the more reaction there is.

Marc Goldring assembled different colors of leather for this box lid. All edges are edge beveled. It helps to define each area as well as play up juxtaposed edges.

A briefcase by Steve Mirones. *Courtesy Steve Mirones*

Edging, Bending, Folding, and Creasing

To finish an edge, many artisans use an edge cutter or sharp knife. Removing a sliver of leather at a very slight angle along the edges chamfers and finishes the edge.

Lightweight leathers may be bent, folded, and creased easily by running a bone folder or inkless ball-point pen along the line with sufficient pressure to leave a mark: Heavier leathers, however (those over 3 ounces), require more preparation. Dampening the crease area will help to maintain a crease. But most often a shallow U-shaped or V-shaped groove is gouged out of the flesh (back) side, so that the material will lay flat and not bunch there. Creases can also be made by pounding with a dull chisel and hammer.

"Squirrel Bag" by William O. Huggins. Leather, fur, bone, abalone, and brass are combined in a unique bag. Drawings on the leather are sketched in ink. *Courtesy William O. Huggins*

Another bag by William O. Huggins in closed position. *Courtesy William O. Huggins*

Joining Parts

Overlapped parts are usually attached by applying a rubber-type leather cement to both surfaces. When this adhesive becomes tacky the parts are joined and pounded together with a mallet.

Leather can also be sewn, if thin enough (garment leather), with a sewing machine or by hand. When hand stitching, it is a good idea to prepunch holes with an awl, thong chisel, or hole punch. Which instrument you use depends upon the thickness of the thread or lacing. Various kinds of stitches can be employed. Some fancier ones can form a decorative knotted pattern along the edge of the seam. To maintain depth of stitching and mark distances between stitches, a line should be ruled, and stitch holes marked with an awl or wheel.

Dyeing

Most often leather is dyed and finished before sewing commences (but after holes are made for stitching, if any). Since undyed leathers are porous, they will absorb dyes readily. Their application should be performed with

This Taureg wallet from Timbuktu, Mali, Africa, is two pieces. The top slides down a braided "cord" fitting like an envelope over the flap-covered compartments of the wallet.

The design is made by slicing and then peeling away thin slivers of the top surface of the leather. Spots of color (marking pen) are added to certain areas bared of the top skin.

even, continuous, rapid, circular strokes. Most commercial dyes for leather are water, oil, spirit soluble. They penetrate deeply into the grain. Holding the dye too long on one spot will cause it to penetrate deeper, imparting a darker uneven color.

To prepare leather for dyeing, make certain the grain side (top) is free of all grease and dirt. Wool, terry towel, and sponge are good dye applicators.

Finishing

The most efficient finishes are those that protect the leather without being obvious. (Dull surfaces such as suedes should not be coated.) Saddle soaps, Vaseline, waxes, and some oils, such as castor oil, make good polishes. Merely rub the finish over the surface, allow it to be absorbed, and then buff off excess.

Laminating and Forming

Although leather comes in limited thicknesses, it can be built into thicker forms by laminating several layers or by shapes forming it into three-dimensional shapes.

Lamination involves adhering of one piece to another with a leather adhesive. After shapes have been stacked and glued, they can be carved and formed with wood-carving tools or with various bits attached to a flexible shaft drill (or regular hand drill).

Forming is accomplished by soaking vegetable-tanned leather until wet, and stretching, shaping, and forming it over some kind of mold. Pine, balsa, or plaster make adequate molds. Collar or saddle four- to eight-ounce leather performs best in this application because it has greater elasticity.

WET-FORMING AND LAMINATING A BOX by Marcia Lloyd

A pattern is made and two triangle pieces are cut for the cover. Saddle soap is applied before the dye. Besides controlling the absorption of dye, saddle soap removes salt left by the tannery. A burnisher is used after soaping to impart a high gloss to the leather. (Smooth stones can be used as well.)

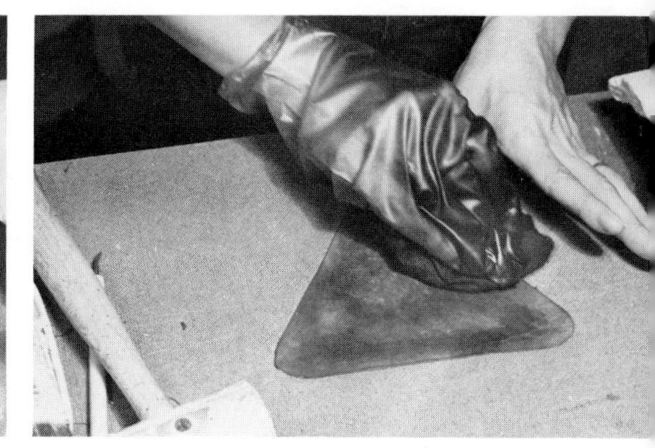

Dye is applied in massaging motions, moving in a continuous circular pattern all over the leather.

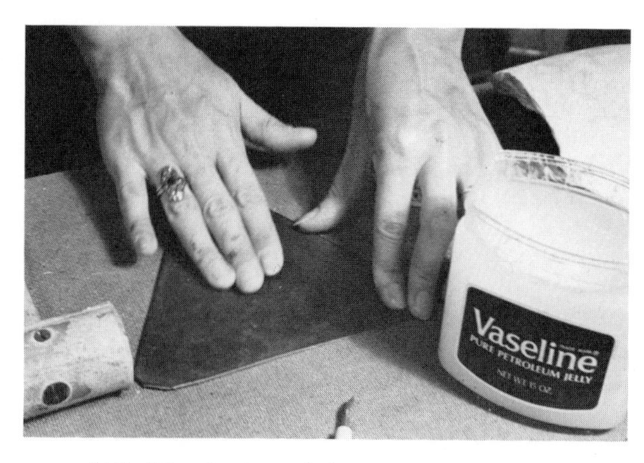

A bit of Vaseline is applied over the dyed surface in a light coat.

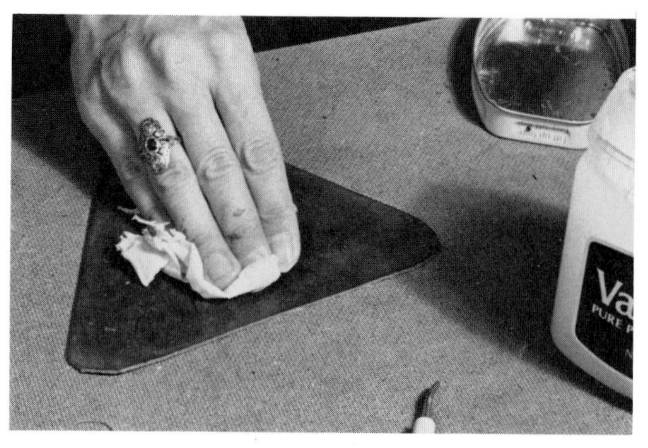

Excess Vaseline is wiped away.

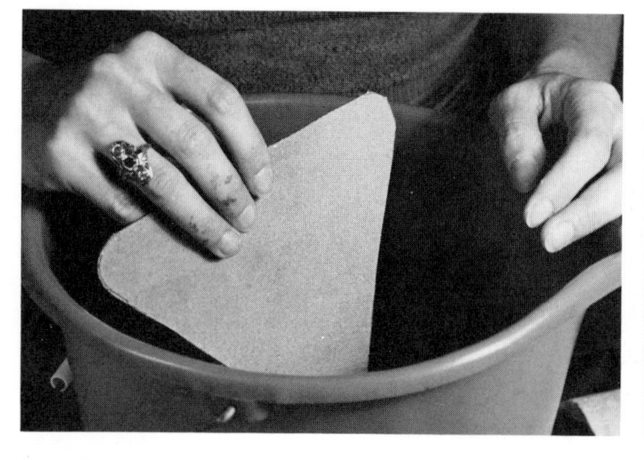

The piece of three-ounce English cowhide is placed into water.

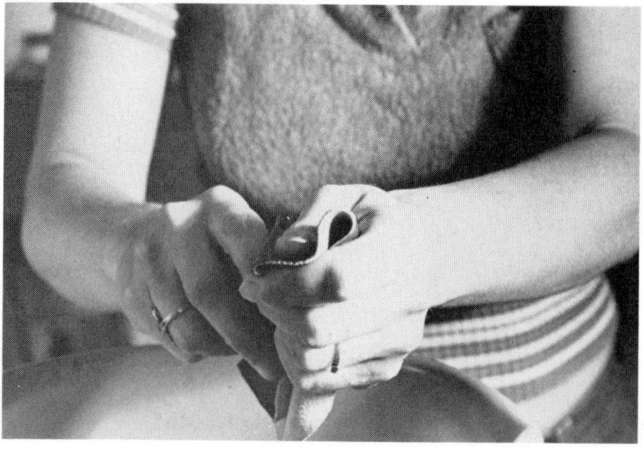

It is left in the water for a moment or two and then removed. It should feel very soft and pliable.

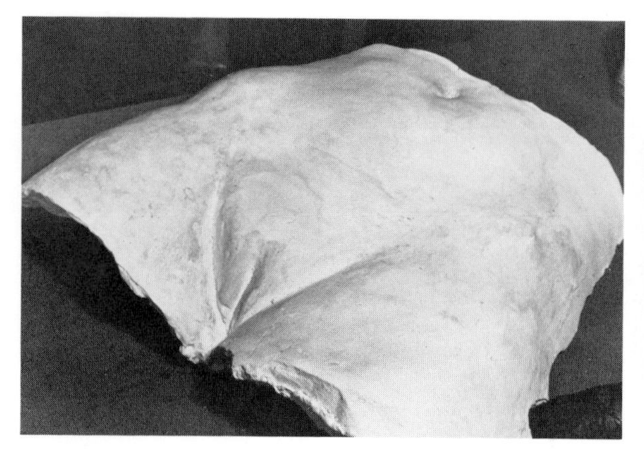

The plaster mold used for forming.

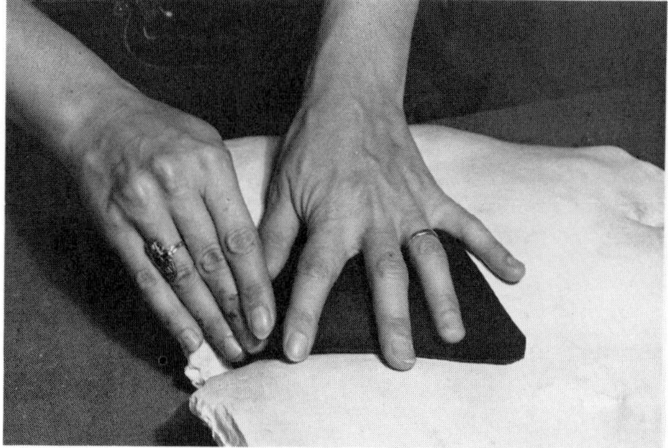

The wet leather is placed on the plaster mold, stretched, and shaped.

Heat from a hair dryer evaporates excess water.

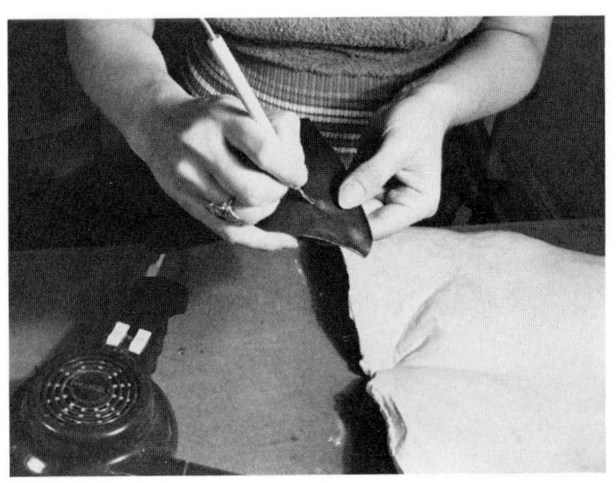

The hair dryer is used intermittently while details are shaped with an embossing tool.

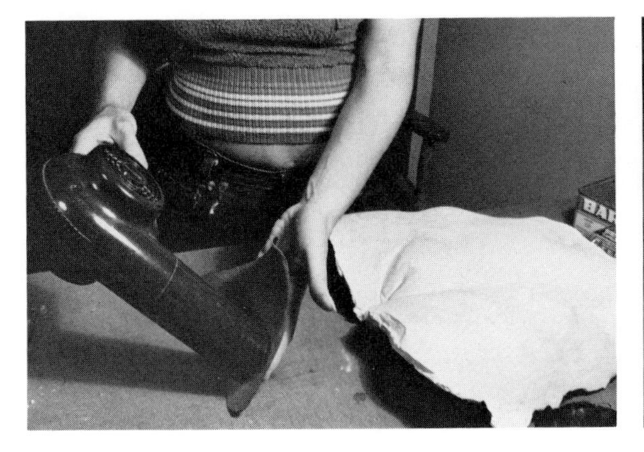

The piece is dried further by Marcia Lloyd. . . .

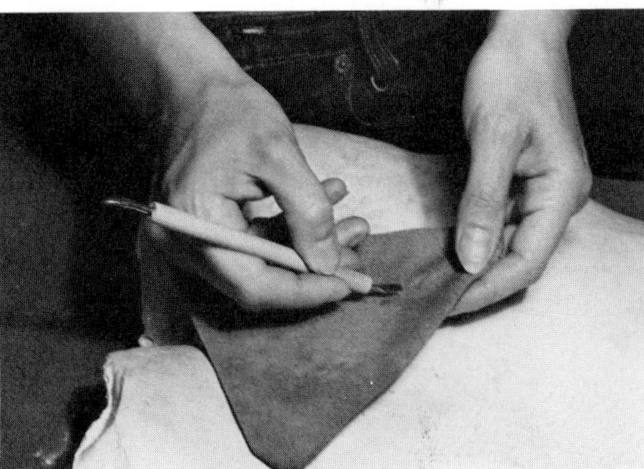

. . . and embossed further.

Edges are sanded.

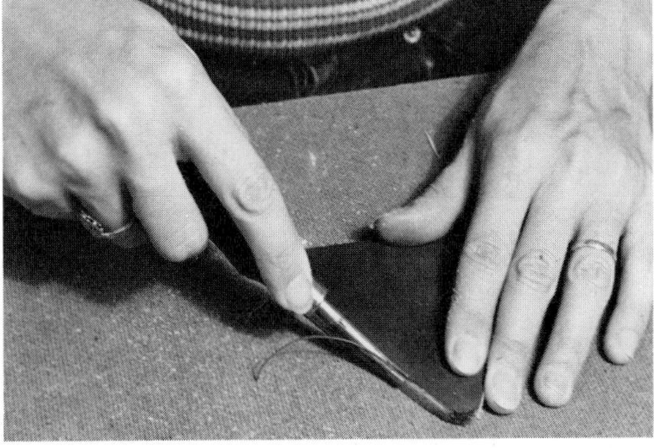

The edge is beveled with an edge beveler.

The edge is dyed with Fiebing's dye.

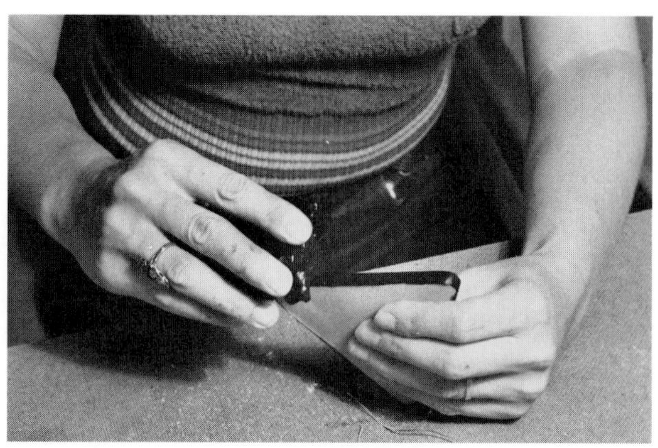

And edging wax is rubbed over the edge to further seal it, minimize, and protect it from future wear.

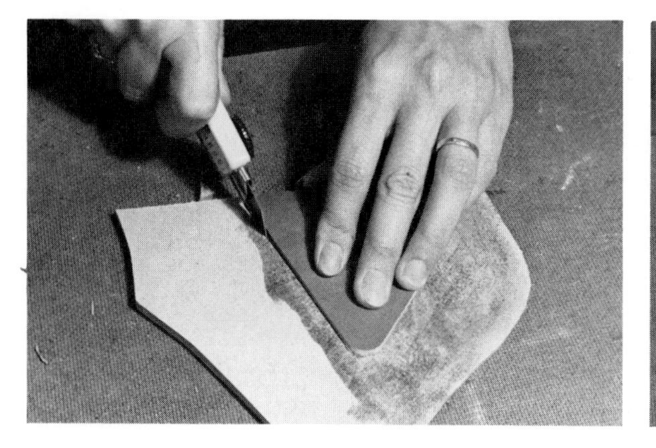

Additional pieces are cut from 9- or 10-ounce English cowhide.

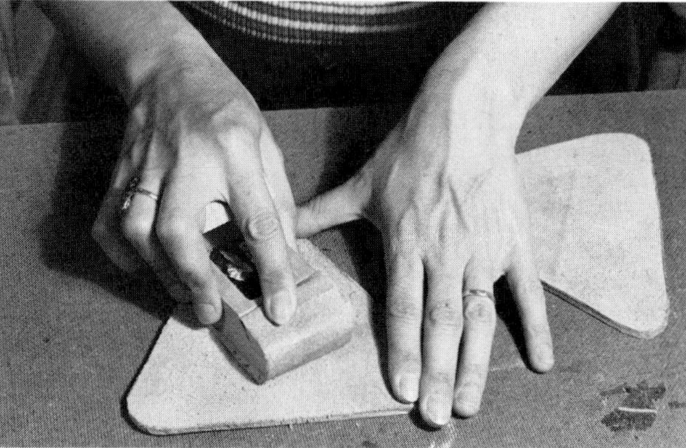

All surfaces to be attached are sanded.

Barge Cement is applied to both surfaces to be adhered and allowed to dry until tacky—for at least 15 minutes.

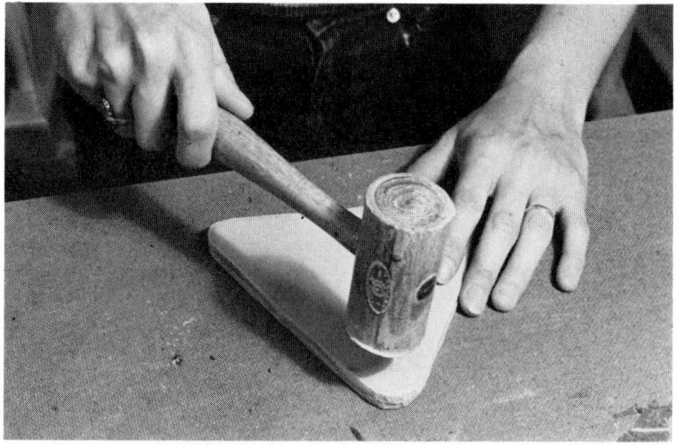

The two parts for the cover are attached. (There are also two for the base and the six pieces for the center). Pounding with a mallet removes trapped air and completes the gluing.

The molded top is placed over the two glued pieces for the box lid.

Because of the shape of the formed top piece, a piece of leather (9- or 10-ounce English cowhide) is cut to fill the gap.

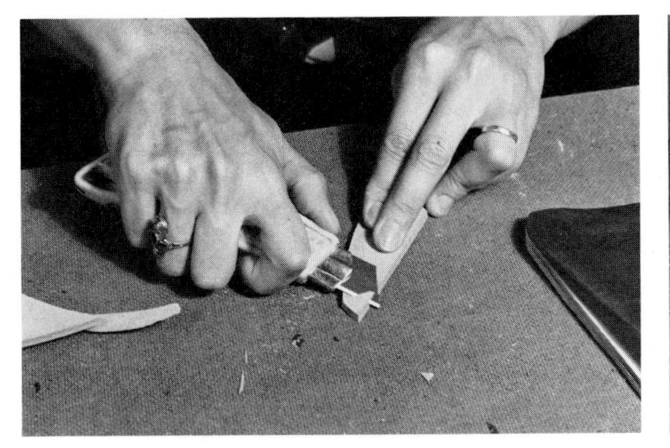

Pieces are carved away so that the filler piece fits precisely. Some flexibility in the leather permits a final adaptation of the leather to the base and fine cracks are sealed.

The filler pieces are glued into place.

A place in the body of the box, where the cavity is to be, is marked, and unwanted pieces are cut away.

A piece of suede is fitted inside the body of the box before any other cutting is done.

The top edge is sanded smooth.

A piece of leather is measured for the cover insert, cut, and tested for fit.

The lid is fitted over the center (body) of the box. Excess leather is cut away.

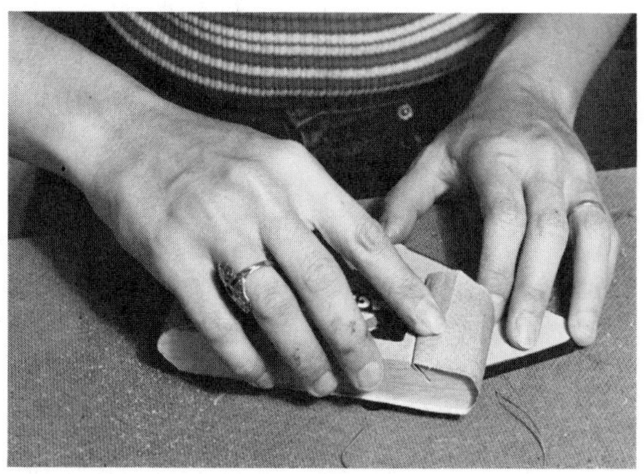

The lid insert and lid are sanded where they are to be adhered. And before adhering the insert, the edge is dyed and sealed.

Barge Cement is brushed on both surfaces, allowed to dry to a tacky state, and the lid insert is compressed with a mallet.

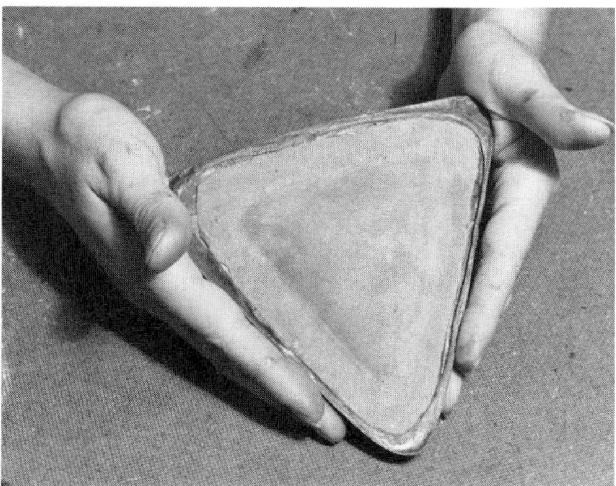

Meanwhile, the base is formed. The area in the center, or cavity, of the box is scribed onto the base with an awl to demarcate where foam padding will go. Foam is cut, glued, and centered. A suede lining is glued over both the foam and base to the edge. The foam creates a soft-raised area in the base.

The three main sections now are ready for assembling.

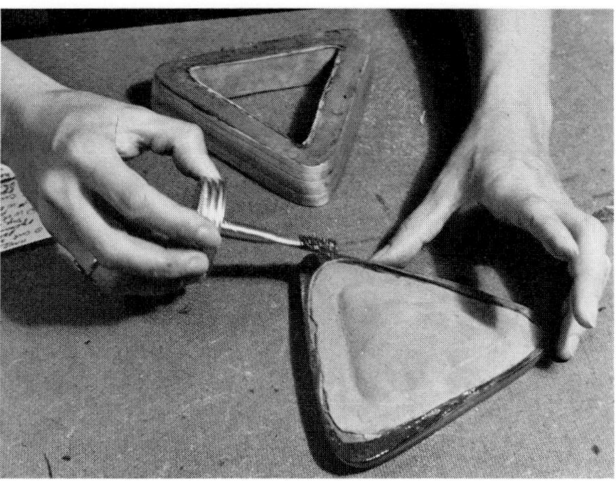

Barge Cement is carefully applied to the edge of the base and the body of the box. Then the two parts are attached and pounded with a mallet.

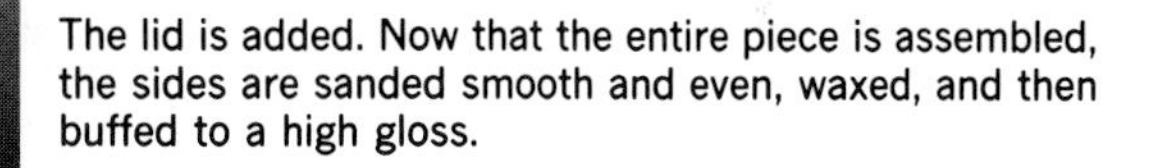

The lid is added. Now that the entire piece is assembled, the sides are sanded smooth and even, waxed, and then buffed to a high gloss.

Marcia Lloyd's "Pandora's Box."

A view of the interior; note the suede lining.

Above: The interior of this box is fitted with dark brown suede.

Right: A series of laminated pots with carved lids by Murry Kusmin.

Below: The front (*right*) and back of a formed bag by Marcia Lloyd. The shoulder strap is hand stitched.

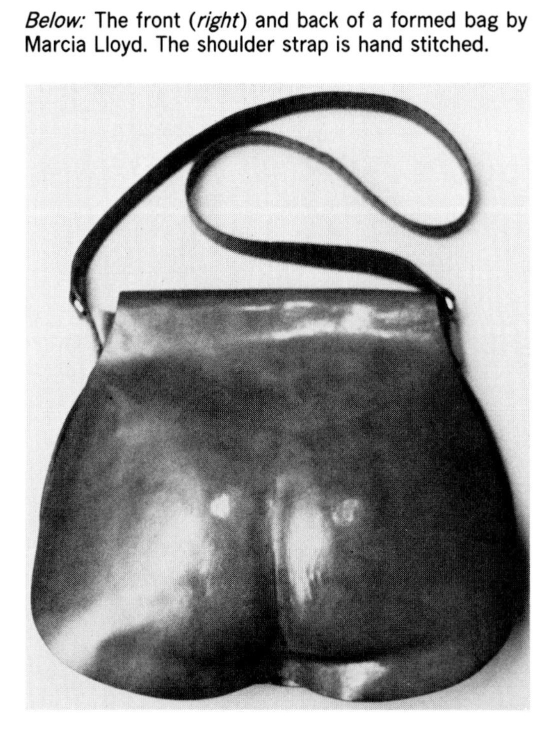

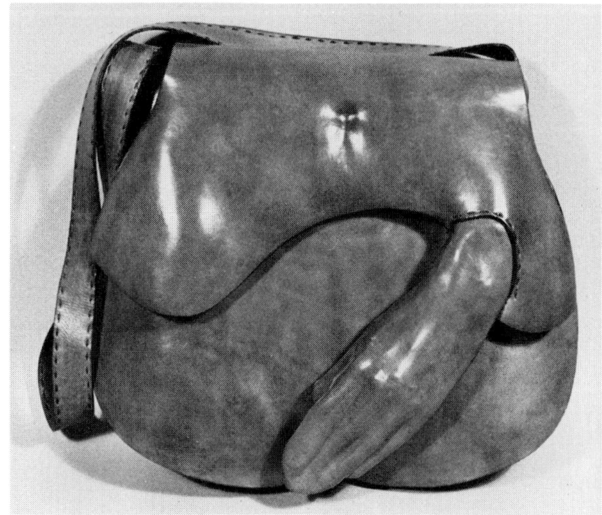

Decorating the Surface

Leather can be decorated by embroidering or weaving on or through it with metal, fibers, threads, plastic, beads, feathers, or more leather. Leather can also be painted with acrylic paints or with leather dyes.

Procedure for Creating Containers of Leather

1. Construct model
2. Make paper or cardboard patterns
3. Trace pattern onto leather
4. Cut leather parts
5. If forming, wet leather and shape over a mold. When dry, trim and refine contours
6. If laminating, glue layers, carve and refine contours
7. If necessary, punch or pierce holes along edges to be joined with thread or thonging
8. Coat edges
9. Decorate and/or dye top or grain side of leather
10. Join parts by gluing and/or sewing or lacing
11. Polish/protect surface

WOOD

Some Words about Wood

Wood is one of the earliest materials used by mankind. It is particularly useful as a container medium becasue it is rigid and dense and yet soft enough to be cut and shaped with simple tools. Because wood is strong and relatively lightweight, it has been hewn into special containers—forms meant to last a long time. Some were more ceremonial than practical.

Treen, the word used to describe small wood objects, were to be found in all areas of living, all over the world. Apothecary jars, bleeding bowls, camphor containers, pillboxes, witches' brew bowls, drinking cups, loving cups, milkmaid cups, Wassail bowls, fruit bowls, platters, salt boxes, spice boxes, mortars, dressing cases, snuffboxes, and flasks are just a few of the ordinary and sublime container forms of yesteryear. Some of the woods were commonplace—pine was and still is, but lignum vitae was considered to be a precious wood and was used with the same reverence as silver.

Wood is a living substance (until the tree is cut down) and should be treated as if the material still lives. In fact, wood cells still contain a small percentage of water for a great many years, even after seasoning (the drying out of green wood). Because of the fact that some water eventually evaporates, wood moves—it shrinks and can warp and crack easily. Each board has its two basic kinds of cuts: crosscut, across the grain, and rip cut, along the grain.

Each kind of wood has its own grain and texture. In fact, each tree has its own particular grain with figure (pattern) determined by the story of the tree —where it lived, what kind of weather there was, catastrophes such as insect invasions, lightning, draft, where the branches were, and so on.

Basically, we utilize wood by cutting it away, building it up by fastening or laminating two or more pieces together, and bending it. The first step in working with wood is a cutting away process, squaring the boards by running them through a planer and then a jointer.

Wood, by virtue of its nature, is a three-dimensional material that has height, width, and depth. Working with wood involves seeing forms on all sides, from all angles—quite different from starting with a flat, hard sheet of metal or a soft length of cloth.

When we utilize any woodworking technique, we are transforming or significantly changing the shape of the wood. Yet we must allow wood to be what it is: wood. Its integrity must never be violated by forcing it into another role.

Working with Wood

Tools

Wood has been around for a long time and all kinds of tools and machinery have been invented to cope with design problems. There are three levels of equipment: hand tools, hand-held machine-powered tools, and stationary, heavy, motor-driven equipment. Even within each area, there are a variety of tools and equipment ranging from hobby instruments to heavy-duty specialized, professional equipment.

We usually start with basic pieces of lumber, cut them into shapes, and work out means for attaching these pieces. Or we begin with a thick block of wood and carve away unwanted areas. The result is called *monolyxous* because the end product was fashioned from a single whole piece of wood and no parts were attached. Bowls are frequently formed this way.

Most woodworking tools, therefore, are designed to cut, drill, gouge, or carve the wood. Even materials, tools, and machines used for finishing slightly cut into the surface while refining it.

Saws, measuring devices, chisels, gouges, mallets, drills, vises, clamps, and planes are among the very basic necessities for most woodworking processes. Specialized kinds of operations require additional tools and equipment. A coping saw is used when cutting fine contours between areas with close tolerances. Special jigs—like the miter box used to cut 45° angles—are useful too. Sharp knives and scissors are helpful tools for trimming veneers. But wood turning requires a motor-driven lathe. In fact, most woodworkers today utilize motor-operated equipment. Time is too valuable to spend using hand-propelled tools, no matter how satisfying the latter may be.

Joining and Gluing Wood

The longest lasting and often the most attractive way to join wood is to cut joints into it and then attach the parts with adhesives. The least satisfactory

A well-aged burl waiting to be carved. *Courtesy Mark Lindquist*

Mark Lindquist carving the burl into a bowl. The horse and tools are of his own design. *Courtesy Mark Lindquist*

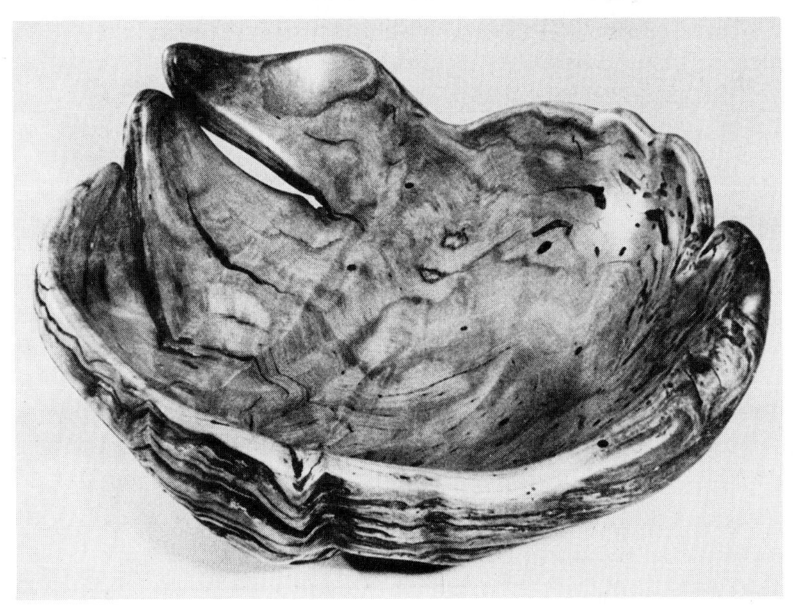

Carved great American chestnut burl container (5" high and 15" diam.) by Mark Lindquist. *Courtesy Mark Lindquist*

way to join hardwoods, which are denser than softwoods and may split, is through use of mechanical fasteners such as nails or screws.

Before any joining is accomplished, ends and sides have to be planed flat or square so that the parts to be assembled fit together without any gaps. After that there are many alternatives. Doweled joints are used to reinforce butt joining in frame-type boxes where the square end of one piece butts up against the flat surface of another. Holes drilled into the butted sections accommodate reinforcing dowels.

When there is a need for mitered corners, feathers, or keys—straight or triangular pieces of wood—are inserted into slots across the mitered joint.

Dovetail joints are another popular and efficient joint for container forms. Zigzaglike negative and positive sections are designed into the wood ends with female areas cut out of each end so that the corners interlock tightly.

Two types of adhesives are most commonly employed to adhere wood joints. One is Weldwood, a plastic resin that comes in powdered form and requires mixing with a small amount of water. The other type is liquid animal or plastic glue in either the brown, cream-colored, or white form. Titebond, Sobo, and Elmer's are some of the most popular. These glues result in nearly invisible joints that are often stronger than the surrounding wood.

Contact adhesives are excellent for joining nonsimilar surfaces such as wood to mirror, metal, or glass, but are too obvious and not strong enough for use in joints. After application of glue to both surfaces, clamping is necessary until the glue sets.

Veneering

Veneering is the process of covering an inexpensive wood or wood-composite surface with rare wood or woods with an attractive grain. Very thin sheets of wood (1/28 inch to 1/20 inch) are adhered to the base form and then clamped until the glue sets. Excess veneer that overhangs the edges can be cut away with a heavy duty pair of scissors.

Wood Finishing

Pine and common types of plywood, when not surfaced with veneer, can be successfully painted. Most hardwoods have an attractive grain and surface that would be a pity to mask with opaque materials. Even pine looks attractive when stained and varnished.

A wide variety of oils, waxes, lacquers, stains, and varnishes are available. Most often some type of surface protection is necessary to prevent wood's porous surface from staining and becoming ingrained with dirt. Perhaps the most attractive coating is a mixture of one part turpentine to two parts boiled linseed oil, applied with a brush or soft rag at room temperature. At least five applications are necesary, each being allowed to penetrate and dry overnight. The final coat can be applied with #0000 steel wool, smoothing the surface while adding the finish.

Transparent coatings such as water-soluble stains, alcohol stains, and oil stains are often used before applying compatible varnishes. Often these varnishes are plastic types such as the polyurethanes that provide tough, durable coatings. Before applying any clear finish, use a tack cloth to remove the dust that would otherwise become entrapped.

When coating containers that will hold food, nonpoisonous finishes are imperative. Melted beeswax that has been thinned with a bit of mineral oil is best. This mixture is applied just at the melting temperature of wax with a soft cloth.

Olive oil and salad oil are also used on kitchen containers. These need to be reapplied more frequently than the other types.

Some special varnishes, compounded for wooden food containers that will be used for alkaline or acid foods, are also available from wood supply sources.

Where abrasion and heavy wear are factors, a great many good plastic-based varnishes are on the market. These provide excellent finishes. They usually are brushed or sprayed directly on the wood. Application procedures for each are described on their containers.

Making Wood Containers

Perhaps the best way to explain construction techniques for making wood boxes and bowls is to bring you through the actual process. The following step-by-step photographs and their captions illustrate in detail some alternatives.

Before beginning the sequences, let us look at a few preliminaries.

Design and Materials

Prior to any construction, plans or designs have to be made. Where needed, templates and jigs have to be cut or constructed to aid and guide in cutting wood to proper dimensions. The entire construction sequence should be thought through to the finishing process. It is quite simple to box yourself in with errors. One way to begin your plans is to list all the parts and sizes of wood that you will need. Then ask yourself: In what order are parts to be cut? What types of joining are best? In what sequence will parts be assembled? What kind of finish will be used? If a box, will it be lined? Answers to these significant questions can help you avoid costly errors.

Many of the operations shown here are performed by machines. Each step, however, can be carried out by hand. It just takes more time. The sequence of steps, however, is basically the same.

A Basic Box

Here is an outline of basic operations in constructing a wooden box. (This list can also serve as a strategy for planning.)

1. Wood cut to approximate dimensions (all measurements checked)
2. Wood is ripsawed along grain and crosscut across grain. (If corners are to be at right angles, this has to be checked.)
3. All sides have to be surfaced (run through planer and jointer to square its planes). Check angles at joints
4. Joints have to be marked and cut. (All templates or jigs used for joint cutting should be on hand.)
5. Box is assembled in trial form before gluing
6. Glue all parts that have to be glued. Clamp all glued areas
7. If lid of the box is a continuation of the body of the box, the solid form you have now should be cut apart, after you mark the depth of the lid
8. If hardware is to be used, such as hinges, then attach at this point (Recess hinges.)
9. Final chamfering (filing off sharp edges with file to form slight angle), planing, sanding takes place
10. Apply finish to box
11. If box is to be lined, construct or cut your lining and glue to box interior

MAKING A BASIC BOX
by Wayne Raab

Boards are cut to approximate size and planed.

A jointer is used to square the edges.

Box sides are marked and the board is cut in two on a band saw so that there are two identical pieces.

Parts are separated on a band saw to make four sides.

In order for the grain pattern of the wood to flow around the box, parts are book matched (and numbered) by flipping over the sides.

Using a production table on a table saw, miters (45° angle) are cut for all corners.

Parts are matched to check for accuracy of cut.

Spline grooves for spline joints are cut in mitered corners and at the bottom for a recessed base. On the bottom, a slight gap is allowed to accommodate expansion of wood, but the splines fit tightly in the corners so that the joints do not separate.

Wayne Raab tests the fit for splines. . . .

. . . and checks alignment.

Splines are cut so that the grain goes in the same direction as the box for strength.

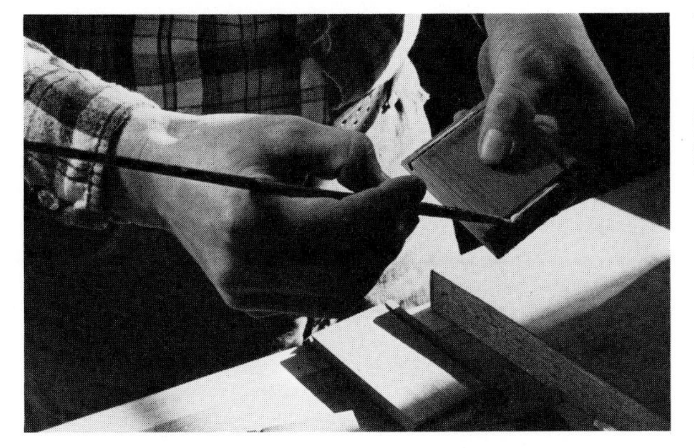

The inside corners of the box are waxed to keep glues that will ooze out of the joints from sticking. Titebond, an aliphatic resin glue, is brushed onto joints and on the first small splines. There is five minutes of working time before the glue sets. Splines are inserted. Corners and base are joined. The base locks into place.

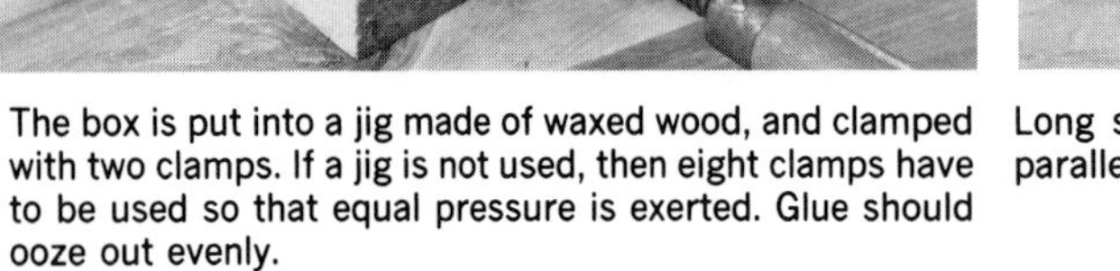

The box is put into a jig made of waxed wood, and clamped with two clamps. If a jig is not used, then eight clamps have to be used so that equal pressure is exerted. Glue should ooze out evenly.

Long splines are glued and inserted. Grain is horizontal, parallel to the grain of the box sides.

Projecting pieces of splines are pared. . . .

. . . then planed, flush with the top edge. The box is supported in a vise.

A rough-cut piece of wood, slightly larger in length and width than the box, is measured for a precisely fitting lid.

A carbide-tipped ripsaw blade is adjusted and the lid is trimmed. A mark is made as to where the lip is to be cut so that part of the lid will sit within the box. That margin of wood is eliminated by routing.

The rosewood lid is shaped with a chisel and hand-held grinder and then the entire form is sanded until smooth. The box is oiled and complete. By Wayne Raab.

Although the base can be produced in multiples, each lid is unique. Walnut boxes by Wayne Raab.

A nonproduction box with curly maple lid. By Wayne Raab.

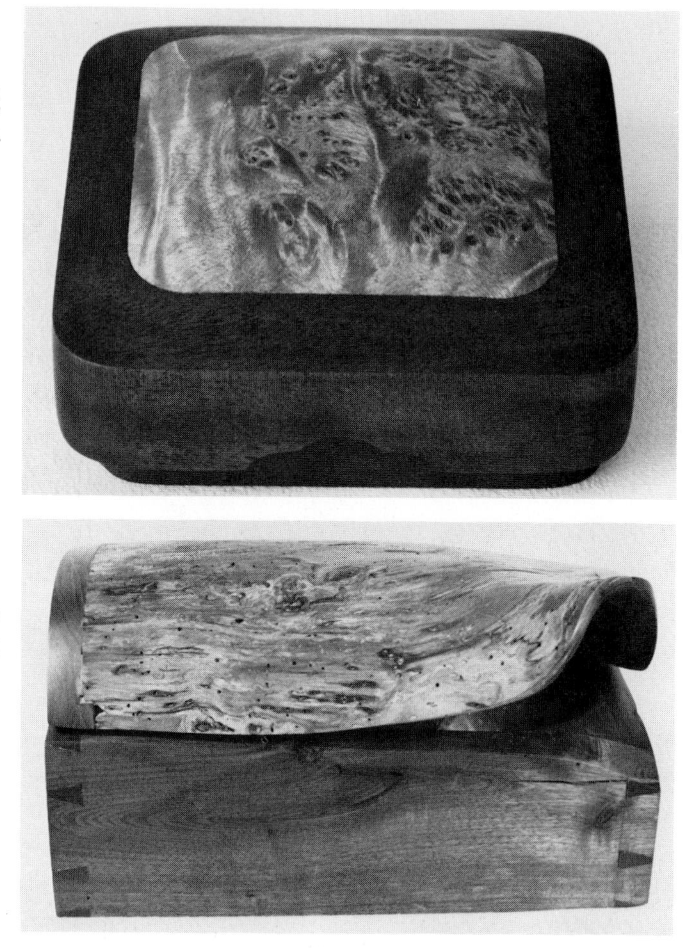

Gary Groves's box of walnut with a maple burl insert (3½" X 3½" X 1½" high). *Courtesy Gary Groves, Woodworks*

The deformations of an apple tree were utilized in this lid. The box is walnut (17½" long, 8½" deep, 8½" high). By Andrew Willner.

A Veneered Box

When veneering a box and when creating marquetry designs in veneer, here is a list of basic operations. The sequence is the same as above from 1 to 6. Some woodworkers shortcut the joining process by gluing and nailing joints. These will be completely covered with veneer later.

1–6. Same as for basic box

7. Glue veneer to each of the sides and clamp in place. (If marquetry is to be used, cut your pieces on a jigsaw, with a hand-coping saw or scissors, beforehand.)
8. Trim excess veneer from around sides
9. Follow above sequence from 7 to 10 as in basic box

MARQUETRY BOXES by Albert Rosenblatt

This series shows Albert Rosenblatt working on two different boxes.

Albert Rosenblatt adjusts his table-saw blade in preparation for cutting the parts of a basic box.

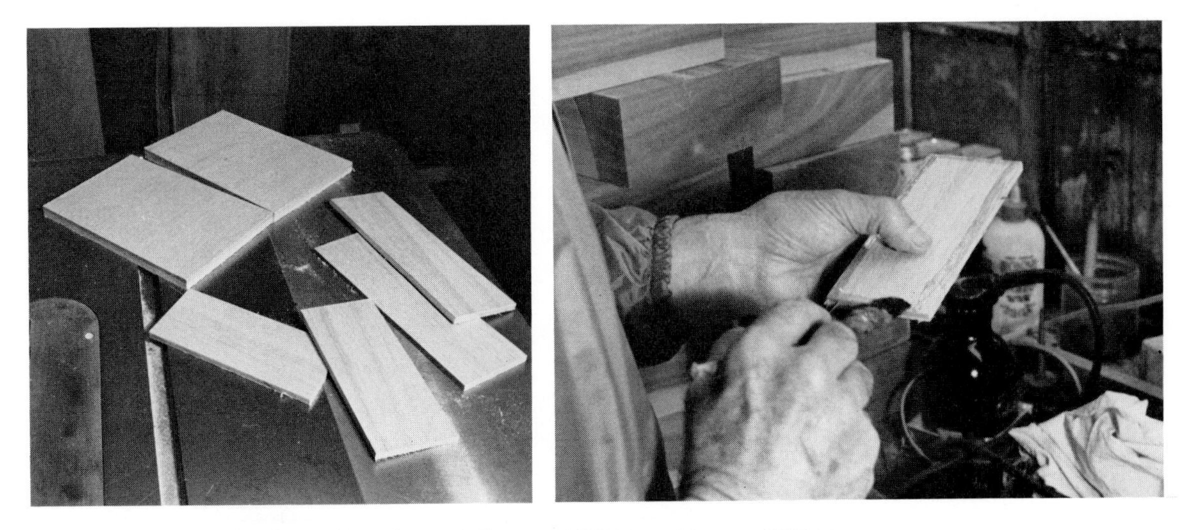

Six pieces of wood are cut. Four are ¼" thick and two are ½" thick. The ends must be thicker because of the method of attaching parts of the box. Philippine mahogany plywood is used. Contact cement (Bondmaster G590, National Starch & Chemical Corp., 10 Finderne Avenue, Bridgewater, New Jersey 08876) is brushed on all edges to be adhered.

The Philippine mahogany side is placed on the inside, since veneer will be used to cover the outside.

It takes 10 to 15 minutes for the cement to become tacky. At that point, the box is assembled and reinforced with two-penny (1" long) brads on the sides.

Three sides and the bottom of this box are covered with mahogany veneer. (This pattern varies from box to box.) Both box and veneer surfaces are brushed with contact cement and pressed together when tacky. Air pockets are forced out with a tool and edges of veneer are trimmed with heavy duty scissors.

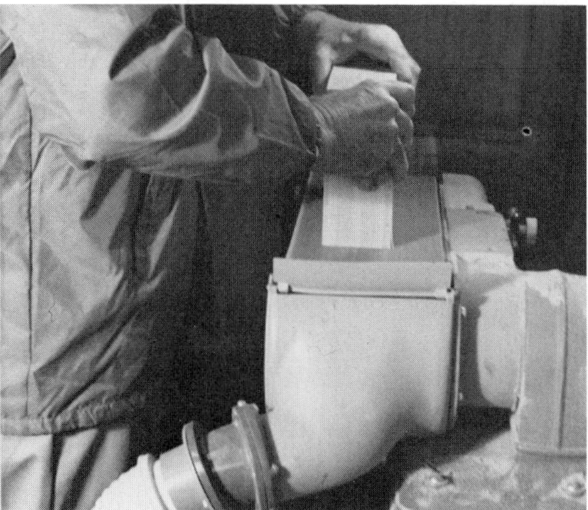

The veneered sides are sanded on a belt sander and the box is ready to receive the two marquetry sides.

To create a marquetry design, a pattern is placed on thin cardboard. Different colors and patterns of veneer are selected and stacked. A sandwich of cardboard is placed on top and bottom. If the veneers are warped, they are dampened with water. The stack of veneers are taped together and placed in a press (made of two thick pieces of wood and clamps).

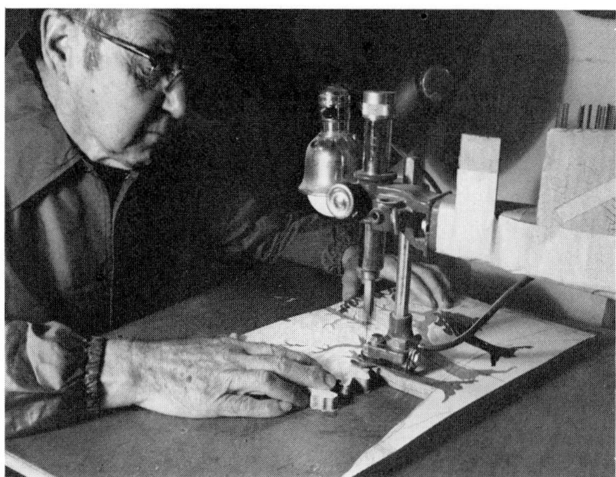

After the veneers are pressed flat, each piece is taped to the next with double-faced masking tape, adhering one to the other. Six to twelve pieces are used for a "book." (There is need for a minimum thickness to aid in cutting.) The pattern or design is taped on top and the whole is cut on a jigsaw fitted with a fine blade. (Less kerf will decrease loss of design and parts then will join closer together.)

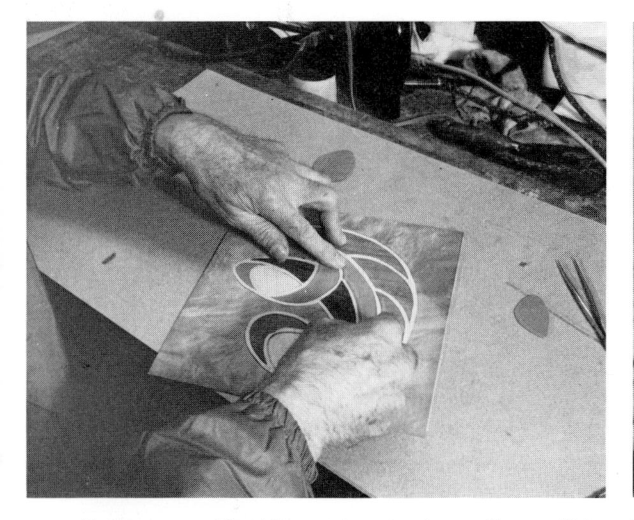

If the veneer "book" contains six pieces, there will be enough cut for six boxes and each will be slightly different. Different colors of veneer from each cut layer are reassembled on masking tape so that pattern areas can be varied.

The marquetry top is measured against the veneered box and the area that extends is used for one side so that the pattern will match and be continuous.

A Kutrimmer paper cutter is used to slice off excess (the side piece).

A generous amount of contact cement is brushed onto both the marquetry and the top and side of the box.

When tacky, contact is made and pressure is applied with a flat tool to the masking-tape-covered marquetry to press out trapped air.

The masking tape is carefully removed after the adhesive has set. Excess is trimmed from the edges with heavy shears.

The box is placed in a vise and sanded with a portable belt sander. Various grits from 120 to 220 are used.

With an orbital sander, edges are chamfered and sides are further refined (to 400 grit).

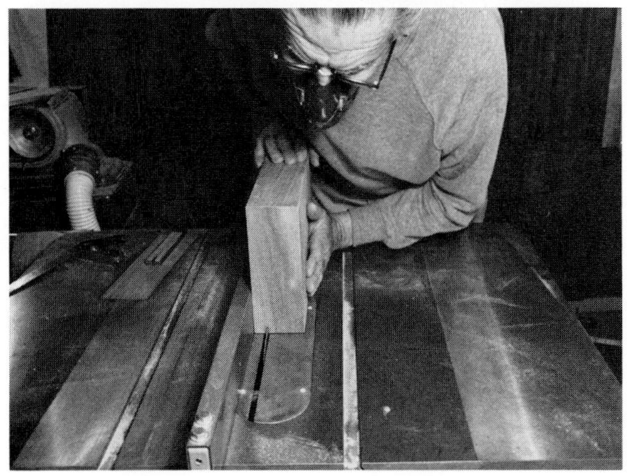

Now the box is opened by setting the table-saw blade to the proper width and cutting apart the box on all sides. A fine-toothed carbide blade (80 teeth to the inch), used for plastic, is employed because it has a shallow kerf—less of the design will be lost, allowing closer match.

A hinge is measured on one side and the hinge area is routed out.

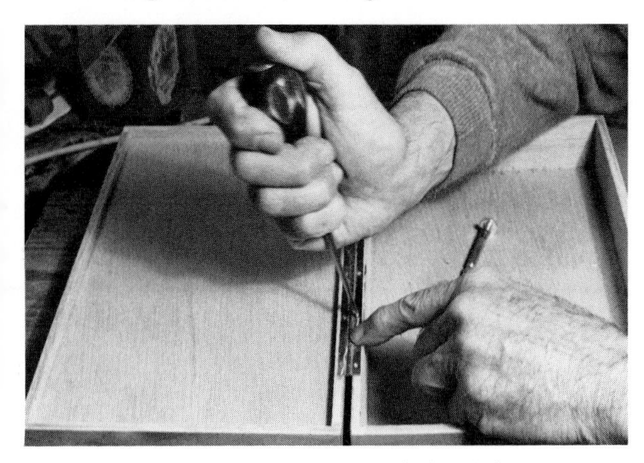

An awl is used to mark for screw holes and . . .

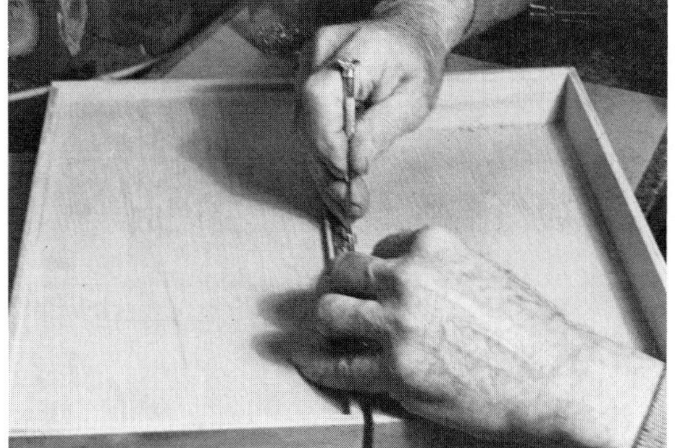

. . . the hinge is screwed in place.

After sanding with 0 steel wool, penetrating sealer is used (704C DuPont Sealer and Penetrating Oil Finish), followed by a coating of Johnson's paste wax. (Some veneers bleed and some sealers cause delamination by dissolving the contact cement.) Two views of a Japanese-style two-tiered marquetry box by Albert Rosenblatt.

Box parts.

Another two-tiered box by Albert Rosenblatt.

Chris Wright's general purpose box. *Courtesy Design Magazine and the Design Council, London; photo by Kokon Chung*

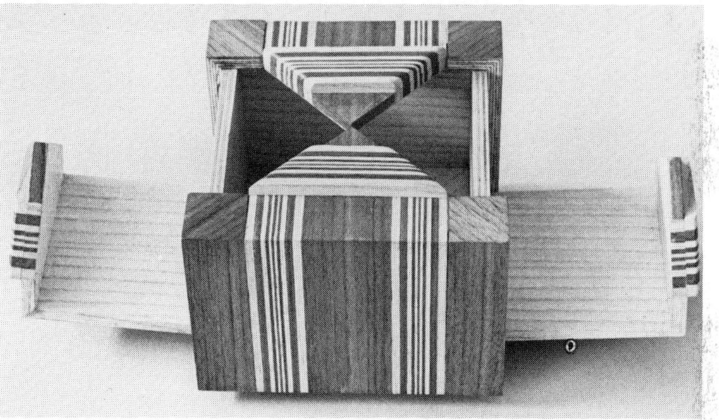

Chris Wright's veneered box opens uniquely as shown in these two views. *Courtesy Design Magazine and the Design Council, London; photo by Kokon Chung*

Veneers wrapped around a heavier base and then painted. From Austria.

Veneer wrapped around a base and colored with marking pens. From Guatemala.

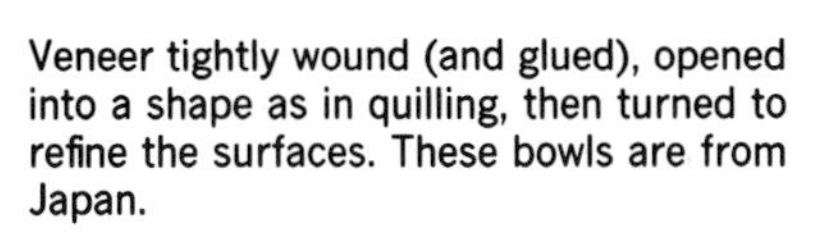

Veneer tightly wound (and glued), opened into a shape as in quilling, then turned to refine the surfaces. These bowls are from Japan.

Turning a Bowl

Wood turning combines the skill of hand-tool work with the power of a machine. A mechanism called a lathe holds the wood while it rotates the form. Shapes are carved into the wood with long handled carving tools as the wood spins. Three basic tools are enough to accomplish most wood-turning operations: a gouge for roughing stock to round shapes; a skew for smooth cutting to finishing of a surface; and a spear-pointed or diamond-shaped tool to part pieces and finish inside surface recesses. Cutting tools must be kept sharp; otherwise they will only scrape (not cut), dig in, and clatter.

The bowl can be turned between two centers (spindle turning) or at the end of the lathe (faceplate turning). To attach a piece of wood to the lathe, locate the center of the block by crossing two diagonal pencil lines from corner to corner. The center is where the lines intersect. Hammer in the spur and attach the wood block to the lathe (for turning between centers). If a plate is to be used, as in faceplate turning, screw the plate into place.

Procedures for Bowl Turning

The following is an outline of the sequence in turning a bowl.

1. Cut roughly shaped block of greenwood
2. Attach to lathe
3. Rough out shape to approximate dimensions so that walls remaining are about one-half inch thick
4. Remove from lathe and place in a warm place where air circulates for seasoning
5. After several weeks to months, reattach to lathe
6. Refine form until final shape is achieved
7. Sand form while on lathe
8. Remove form from lathe and apply finish

BOWL TURNING by Bob Stocksdale

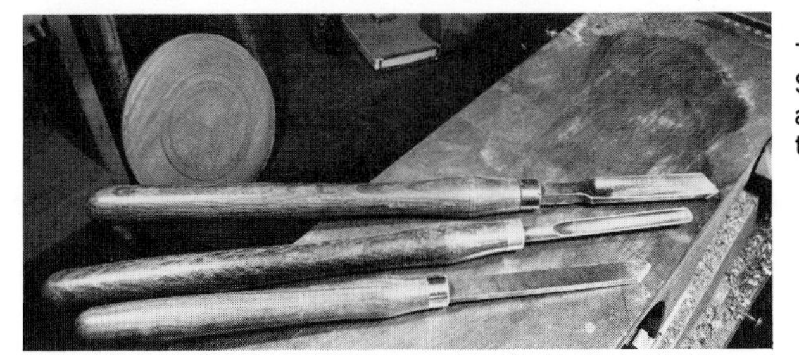

Three basic tools used by Bob Stocksdale are 1" and ½" gouges and a diamond-point skew, handmade of tool steel.

The wheel turns at 1800 rpm for a 9" bowl. Here a diamond-point skew is poised on the rest.

It is used to cut the foot at a slight angle so that a universal chuck can be used to grip the base for later turning.

For the finishing operation, the lathe must be slowed down to 600 rpm. Garnet 36 grit hand-held sandpaper is used. Otherwise, the sandpaper just slides over the work. Every once in a while, rotation is reversed to remove tool and sandpaper marks. Then 50 grit, 100, 150, and finally 220 grit papers are used.

The whole bowl is sanded before finishing the little foot that is clamped in the universal chuck.

Bob Stocksdale rough-turns a bowl to a ½" to ¾" thickness and allows it to stand about one month in a drying room at 90°F. Sometimes the bowl is placed in a plastic bag that is slightly open for a few weeks before placing it in the drying room. That way he can see how the wood reacts. On the left is a rough-turned bowl; on the right, a completed counterpart.

Walnut bowl by Bob Stocksdale.

An olive wood bowl by Bob Stocksdale.

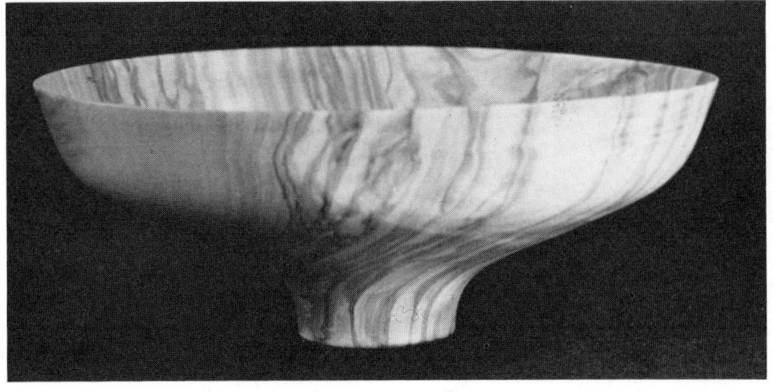

FITTING A WOOD JAR WITH A LID ON THE LATHE by Mark Lindquist

A piece of well-aged burl is band sawed to a cylinder-shaped block. The piece must be relatively free of faults and blemishes. A faceplate is mounted on the piece using sheet-metal screws. One-third of the cylinder is then sliced off on a band saw and that part of the lid is also faceplate mounted. The first (⅔) section is turned and the bowl is rough-sanded inside and out so that the mouth is defined. Then the lid is mounted on the lathe and rough-turned.

Then the lid is more precisely turned and sanded so that it fits the mouth of the container and a lip is formed.

The bowl is remounted on the lathe. The lid is placed on the bowl and the two are held firmly in place together by a live or ball-bearing center (the tailstock of the lathe).

The exterior form of the lid is turned and sanded.

Now the lid and bowl are separated and the lid is remounted on the lathe. The inside is turned and the lid is completed by sanding except for the part attached to the faceplate. The bowl is remounted and the faceplate is removed from the lid. It is replaced in the mouth of the bowl, and a piece of paper, acting as a shim, holds it firmly. The stub (from the faceplate) is turned and sanded and the lid piece is finished by sanding. The lid is removed once again.

The bowl is finish sanded to 400 grit, the faceplate removed, and the bottom finished and sanded. The entire piece is oiled, allowed to dry for a week, and later buffed with a waxy compound. Bowl by Mark Lindquist.
Photos by Mark Lindquist

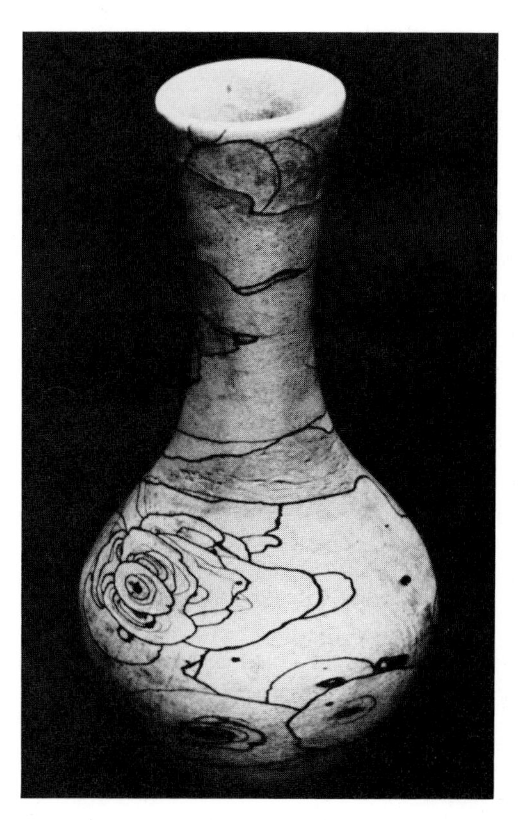

Spalted maple vase lathe-turned by Melvin Lindquist (6" high X 3" diam.). *Photo by Bob LaPree*

Other Woodworking Processes

Bowls can also be carved by hand using Surform tools, chisels, and gouges. Even boxes can be formed that way. Another process is to construct a container form using a solid wood for base and lid and wrapping this with a thicker variety of veneer. Strips of veneer can also be wrapped tightly around a solid piece of wood into a coil while each surface is glued. Any gaps can be filled later with a sawdust–white-glue mixture or with polyester resin. And the surface can be refined either on the lathe or using Surform tools or chisels and gouges.

Hand-carved giant elm burl container/-sculpture chair. "Goliath's Helmet," 3' diameter by Mark Lindquist. *Photo by Mark Lindquist*

Far left: Art (Espenet) Carpenter's band-saw box in walnut.

Left: A slice is taken off the back of the box and areas are cut out. (That sliced-off piece is later glued back.)

Above: The piece that was cut out is then cut again, 90°, and the two parts are reassembled to create a drawer.

Below: Natural tree form carved into a container by Raymond Pelton. *Bottom:* A detail of the above with the drawer pulled out. *Courtesy Raymond Pelton*

Right: Another natural container form by Raymond Pelton. *Bottom:* A "cover" lifts off a tubelike cavity. By Raymond Pelton.

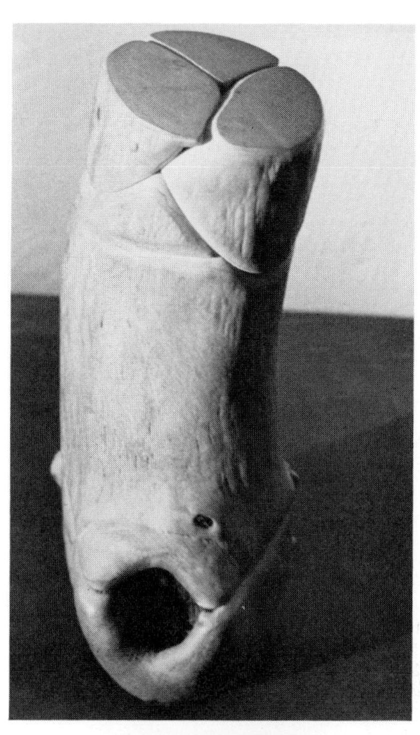

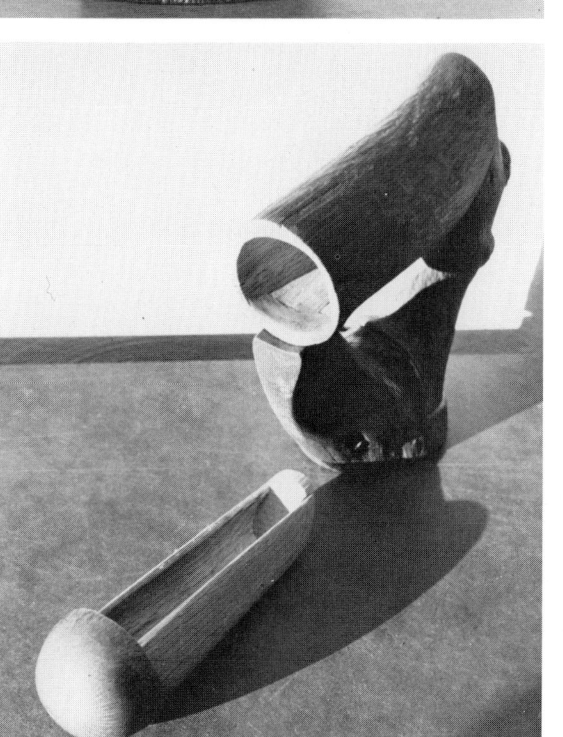

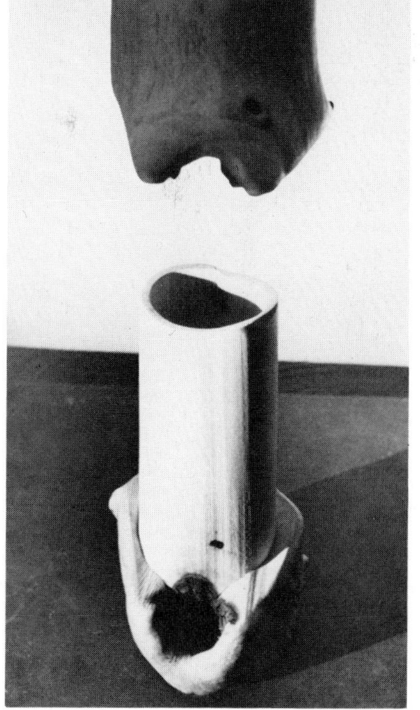

LINING A BOX

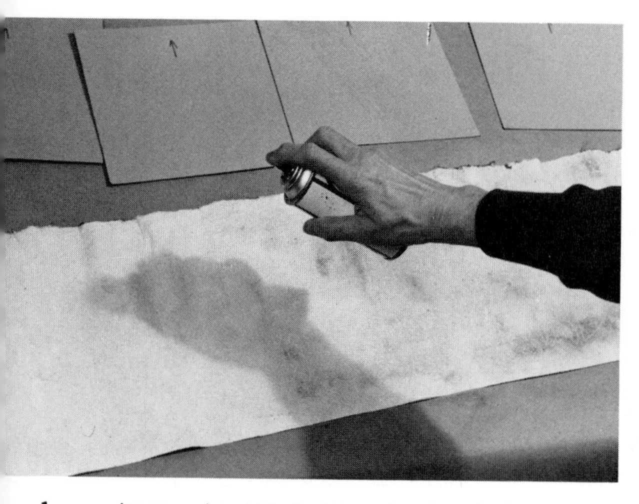

1 Japanese hand-blocked paper is cut to size in a continuous strip to accommodate four pieces of cardboard that would fit the interior of the box. Allowance is made for the thickness of paper that will be used to cover the cardboard. Rubber cement is sprayed (or brushed) on all surfaces to be adhered.

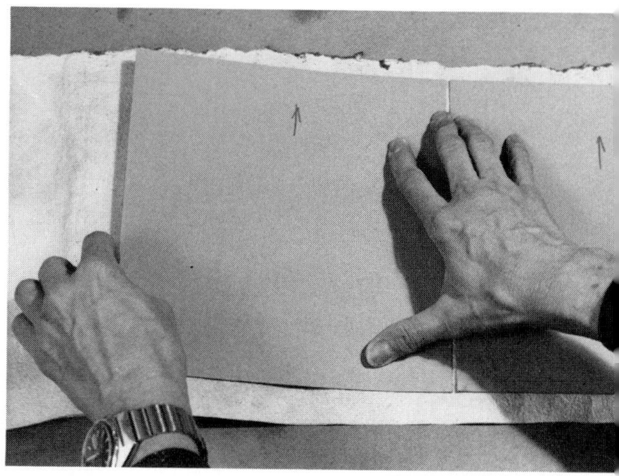

When tacky, the cardboard is pressed in place, leaving room between the pieces for folding of corners. 2

3 A brayer is used to effect a bubble-free bond between paper and backing.

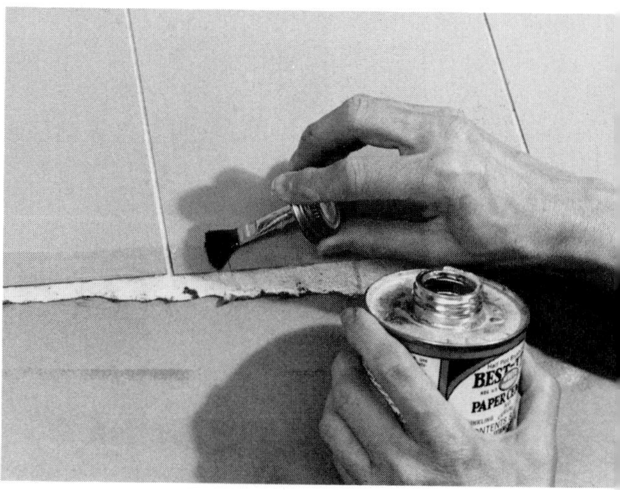

Rubber cement is brushed along the edges of paper and cardboard . . . 4

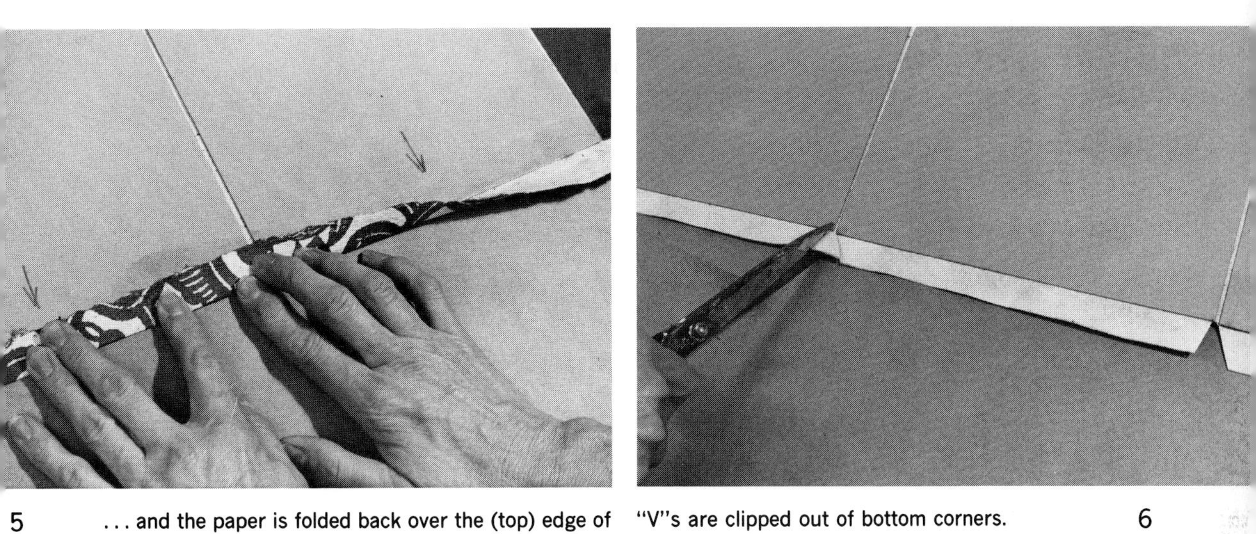

5 . . . and the paper is folded back over the (top) edge of the cardboard.

"V"s are clipped out of bottom corners. 6

7 The bottom piece of cardboard is glued on its own piece of decorative paper.

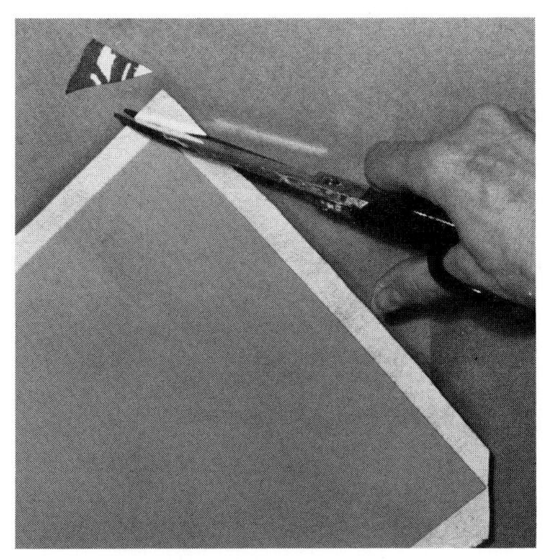

Corners are snipped off to facilitate folding them around the edges of the cardboard. 8

9

Sobo, Elmer's, or a similar white glue is brushed on the back of the cardboard to adhere it to the wood. Note that the bottom edge has not been folded back; rather, it will be bent forward. (The bottom cardboard will rest on these edges.)

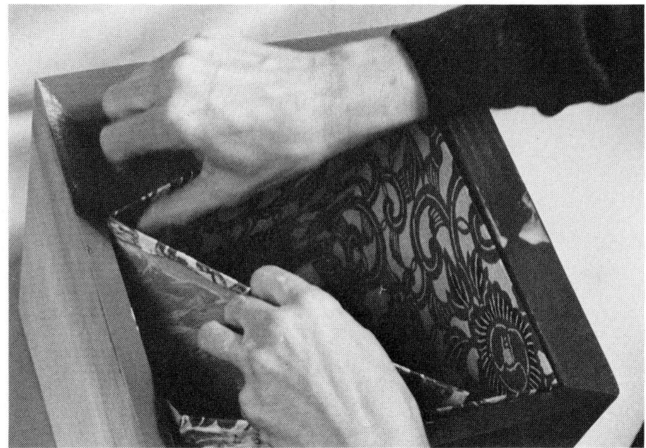

10

The sides are glued into place. Extra glue is brushed along the top edge. Later, clip-on clamps will hold the top edge in place until the adhesive dries.

11

White glue is brushed on the underside of the bottom liner and that is also pressed in place at the bottom of the box. Note the bottom edges of lining paper lie at the box bottom.

A lined "block" box. By the authors.

MAKING PLASTIC CONTAINERS 7

Modern plastics, in thirty years, have taken their place beside many ancient materials—wood, glass, metal, and clay. But while their earliest applications involved the ability to copy the characteristics and appearances of natural materials, today's plastics have shed that imitative image to emerge as an incredible variety of substances that not only do some old tasks better, but fill many new roles as well.

Plastics have unique qualities. Many are easy to work with. Some can be cast into intricate forms, large and small, without elaborate equipment. They have their own textures and working characteristics. For the container-maker, plastics offer enormous potential to develop forms conveniently, inexpensively, and quickly, without any sacrifice in quality or design. As the name implies, plastics are responsive and versatile. With a few basic skills, their potential is open to everyone.

Sheet Plastic: Acrylic

Acrylic remains the best known and most widely used plastic—and with good reason. Although it weighs approximately half as much as glass, it is more flexible, just as optically clear, much more readily worked, and available in a wide range of thicknesses, colors, and textures.

Acrylic sheet is manufactured from methylmethacrylate resin under heat and pressure. In addition to sheets in a standard 4' X 8' size and in thicknesses ranging from 1/16" to several inches, acrylic is available in tubes, rods, balls, custom shapes, and large blocks. Clear acrylic, because it breaks less easily than glass, has been used for years in airplanes, school windows, recreational vehicles, toys, and household appliances. It is also widely available in textures and opaque, translucent, and transparent colors. The latest innovation is acrylic mirror—in several metallic tones. Specialty products, like acrylic cylinders with a mirrored finish—Mirron tubing—are gaining strong footholds too.

The reason for this variety is popularity, and part of the reason for popularity is the ease with which acrylic can be used. In addition to an indefinite shelf life, acrylic can be sawed, drilled, tapped, sanded, polished, machined, cemented, etched, and carved. It can be heated and shaped while it is hot, and returned to its original shape by reheating later.

Acrylic Processes

Acrylic may be machined just as metal and wood are, but it does have properties peculiar to itself alone. Gluing, for example, is usually accomplished with a solvent cement. And in addition to common machine processes, acrylic may be bent along a line by strip heating, or the entire piece may be shaped by heating it in an ordinary oven.

Marking and Sawing

Acrylic sheet comes with a protective paper or plastic covering that is easily peeled away. Whenever possible, it should be allowed to remain on the plastic during machining to avoid unnecessary scratching. The acrylic may be marked for cutting on this covering. If it is necessary to make lines directly on the acrylic, use a grease pencil since it can be wiped away with a soft cloth afterward.

Acrylic sheets, rods, tubes, and blocks are easily cut with any saw made for woodworking or metalworking. When using power equipment, the plastic should be fed through the saw slowly to avoid the gumming that occurs when excessive heat builds up. For straight cutting, circular saws and radial arm saws work well. Jig or saber saws are effective for cutting small curves and intricate patterns. Band saws are effective for most purposes.

Metal-cutting blades are best for acrylic, if no blade specially designed for use on plastics is available. They deliver the cleanest, sharpest cuts, and where possible, they should be carbide-tipped. As with any other machining operation, proper shop precautions should always be taken, including the wearing of protective glasses.

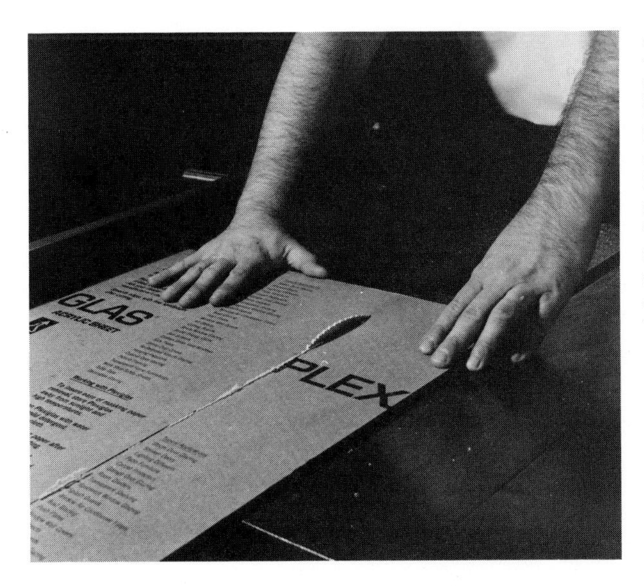

Acrylics may be cut on machinery used for wood or metal. Here a table saw is being used, but band saws (10 teeth/ inch), saber saws, and jigsaws (14 teeth/ inch) may also be used. For table saws, choose a blade made especially for acrylic or one with at least six teeth per inch, all of the same shape, height, and point to point distance. Set the blade slightly above the thickness of the plastic to avoid chipping. No matter what type of saw is employed, hold the acrylic down firmly and do not force feed. *Courtesy Rohm and Haas Company*

Drilling and Tapping

Drill bits specially made for acrylic cut more cleanly than bits for wood or metal. Although they are a recent innovation, many sizes are available. The main feature is a zero rake angle that prevents the bit from catching the plastic. When drilling or tapping, mild soapy water or an oil coolant are helpful in reducing friction.

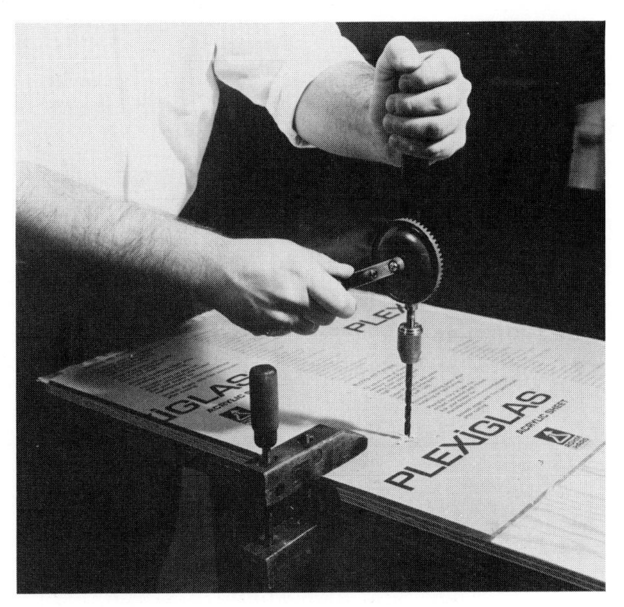

Manual drills, electric drills, and drill presses may all be used effectively and easily with acrylics. The plastic should always be backed with wood and clamped or held securely. Manual drills may be used with metal bits. However, specially ground bits should be used with power tools. Use the highest speed up to 3000 rpm for holes less than ⅜ inch in diameter. For larger holes, lower the speed to 1000–2000 rpm. Never force feed; too much pressure will cause chipping to occur at the back of the sheet. *Courtesy Rohm and Haas Company*

Scraping and Sanding

Scratches and dents in the edges of acrylic sheet can be readily removed by scraping and sanding. The result is known as a "dressed edge." The surface is first scraped with a sharp woodworking scraper. This is like planing. Scraping removes deep scratches that result from manufacturing or machining processes. After scraping, the acrylic should be wet-sanded (with wet or dry sandpaper), using a progression from coarse to fine grits, typically 150, 220, and 400 sandpapers. Sanding completes the dressing, which leaves a matte, translucent surface.

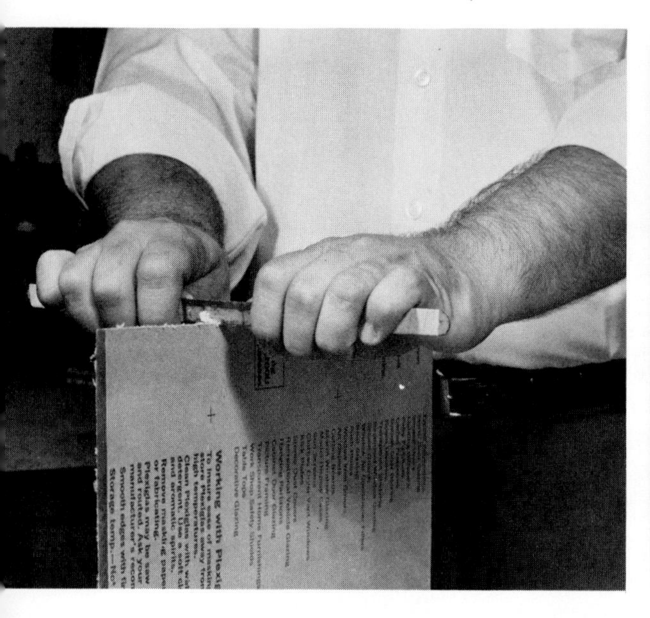

There are three stages to edge finishing: smooth finish, satin finish, and transparent finish. Saw marks and tool marks may be removed with a metal scraper made from a sharpened piece of metal such as the back of a hacksaw blade. They should be removed to guard against breakage due to stresses. Corners may be rounded and uneven areas smoothed with medium to fine tooth metal files. *Courtesy Rohm and Haas Company*

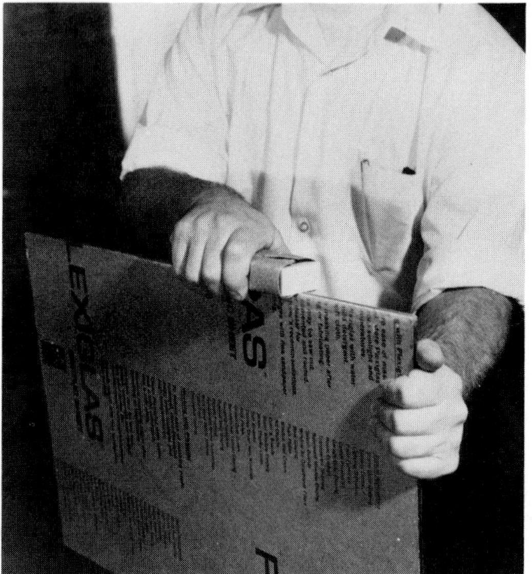

The second stage of finishing, satin finish, improves the appearance of the edge and prepares it for cementing. The edge is sanded with increasingly finer grits of wet or dry sandpaper. Avoid rounding the edges; it produces bubbles in the cemented joint. *Courtesy Rohm and Haas Company*

Polishing and Buffing

A highly glossy, transparent edge can be achieved by polishing and buffing the surface. The best system involves a two-wheel buffer with fairly loose, soft, 10-inch wheels, but buffing wheels attached to hand drills work very well too. In any case, buffers should rotate at no more than 2000 rpm per minute.

Once deep scratches have been scraped out and the edge wet-sanded with different grits down to 400 fine sandpaper, the edge is prepared for buffing. Stick wax should be applied to the buff first to hold the compound to the wheel, and then white tripoli compound is applied.

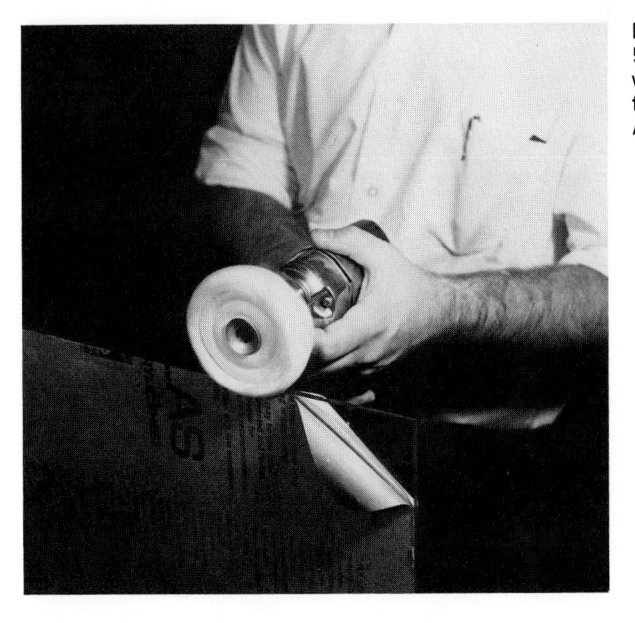

For a glossy edge, continue sanding up to 500 grit wet or dry sandpaper, and buff with a clean muslin cloth or wheel using fine grit buffing compound. *Courtesy Rohm and Haas Company*

Etching and Carving

Acrylic can be etched and carved too. Sharp pointed tools like awls may be used to scratch designs into the surface. Fine drill bits may be used powered with flexible shaft drills to carve and decorate the surface as well. Carbide-tipped tools are the most efficient cutters for acrylic.

Cementing

Once the acrylic has been polished and decorated as desired, pieces may be combined into containers. A strong, almost invisible, bond can be made between pieces of acrylic. Some of the best results may be achieved quite easily using a solvent cement like methylene chloride. The surfaces to be bonded should have either polished or dressed edges. Surfaces are first placed securely together and taped in place as shown on next page. The solvent cement is then applied along the entire length of the joint. A brush or plastic hypodermiclike needle made especially for this purpose may be used to introduce cement to the area to be bonded; capillary action draws the cement into the joint although it is actually applied to only one edge. Solvent cement dries enough to handle most joints in three to ten minutes. It is machine safe in three to four hours. The bond is a powerful one, which reaches full strength after about a day.

Another cement is PS-30. It is a two-part system—resin and catalyst must be mixed. Although it requires slightly more effort to use, PS-30 provides the strongest bond possible from a cement and should be used in high stress situations and where acrylic will be in contact with water.

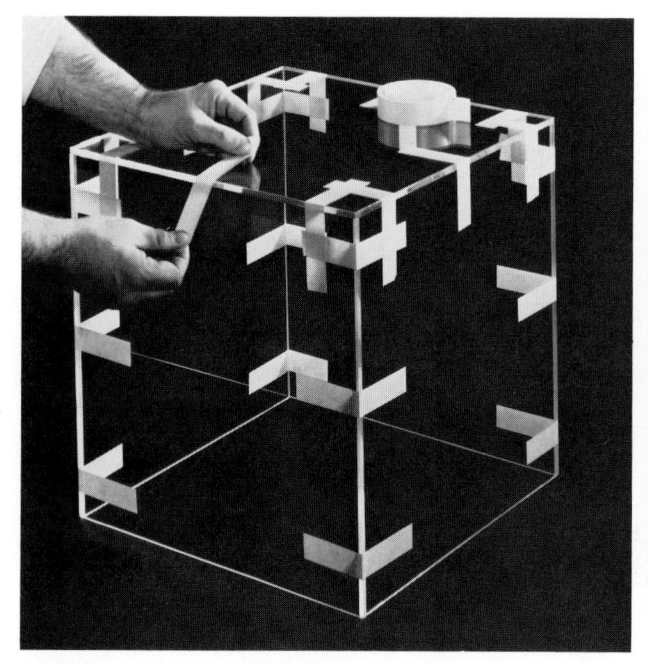

After scraping and sanding the edges to be glued—and buffing the edges that will be exposed—the acrylic form is taped together with masking tape. *Courtesy Rohm and Haas Company*

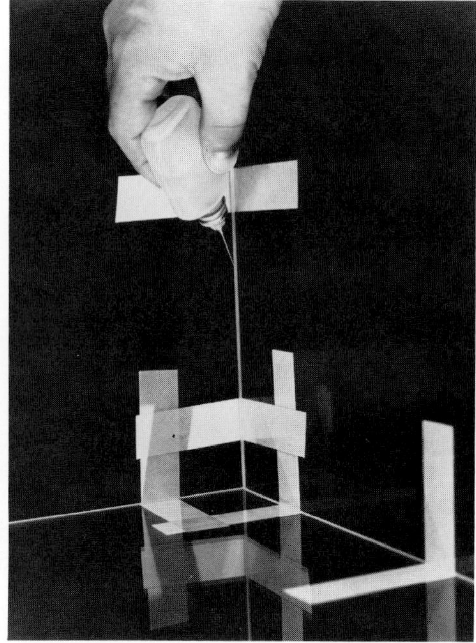

Solvent cement is applied to the joint from a squeeze bottle with a needle-fine spout. The liquid cement flows into the joint by capillary action. *Courtesy Rohm and Haas Company*

Heating and Forming

Acrylic is a thermoplastic—it can be heated and shaped and reheated to return it to its original shape. This property can be capitalized upon very profitably in container-making. Acrylic sheets become flexible, rubberlike, and soft after ten to fifteen minutes at a temperature between 240°F. and 340°F. Once softened, they can be readily formed into three-dimensional shapes. The shape is held until the plastic cools. Once cool, the shape created while the material was hot will be retained.

A regular kitchen oven can be used to heat larger pieces of acrylic. Strip heaters can be used to heat acrylic along a straight line for bending. Be careful with hot acrylic because it is *hot.* Always wear oven mitts when handling hot materials, especially since it may be necessary to hold the acrylic in place until it cools enough to maintain its new shape.

When heating the acrylic in an oven, it is wise to place it on a flat aluminum sheet so that it does not drape itself into the ridges on the rack. Never exceed the recommended temperature range of 240°F. to 340°F. Although acrylic does not ignite until it reaches 700°F., if it is exposed to excessive heat, or is allowed to remain too close to a heating element for too long, it will begin to scorch and bubble. When softened by heat, acrylic may be draped into or over forms and held in place with weights or gloved hands or light pressure from clamps. If the plastic is too springy, it may require additional heating. Cooling can be speeded by placing the acrylic under warm water.

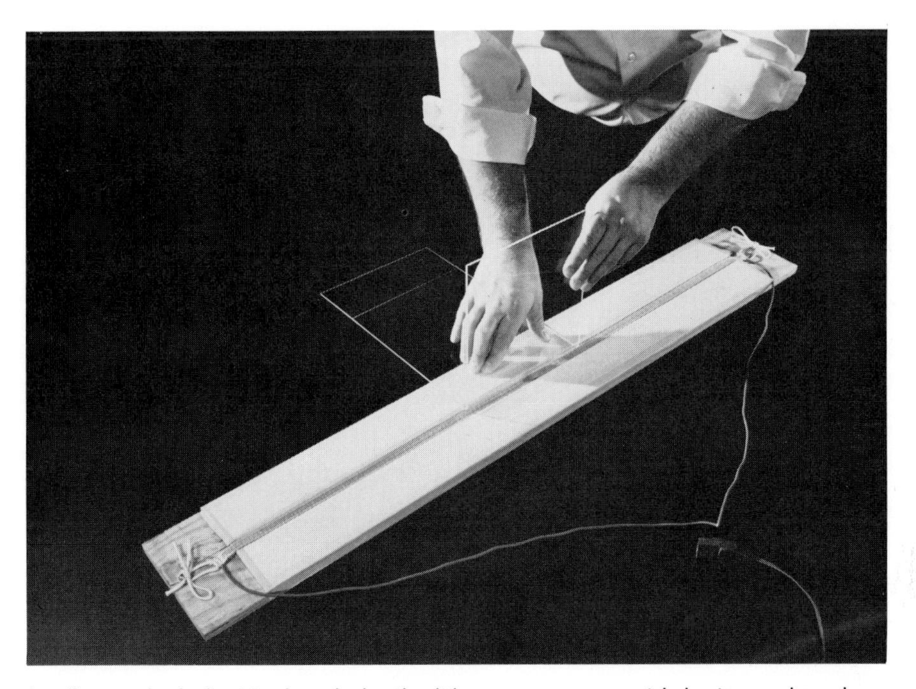

Acrylic may also be bent to shape by heating it in an oven or over a strip heater, as shown here. The strip heater softens the acrylic along a line for easy and accurate bending. The Briskeat RH-36 heating element used here is widely available. *Courtesy Rohm and Haas Company*

CONTAINERS OF ACRYLIC TUBE

Containers can easily be constructed from any length of acrylic tube. Here, Mirron® tubing (Thermoplastic Processes Inc., Valley Road, Stirling, New Jersey 07980), an opaque acrylic tubing with a mirrored outer surface, is being scraped and sanded in preparation for gluing.

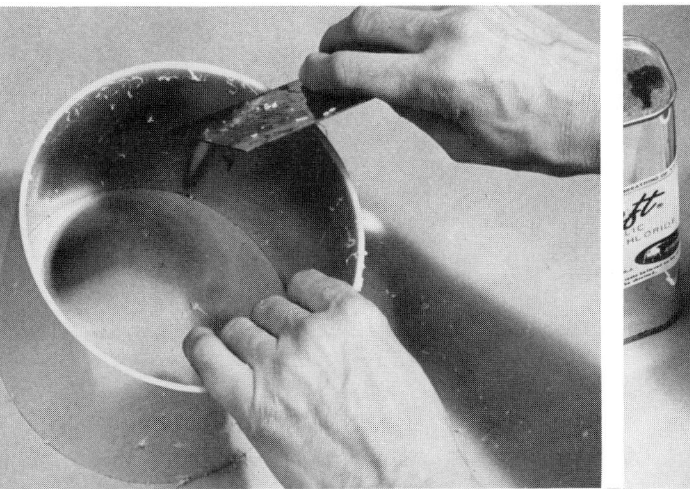

A round piece of clear acrylic was sanded, polished, and taped to the section of tube. Solvent cement is introduced into the joint by a plastic syringe with a flexible plastic needle.

After the cement cures, the tape is removed. A dilute mixture of white glue and water is painted onto the inside of the container (but not on the bottom) as an adhesive for the feltlike lining.

Suede-Tex® powder creates a soft, warm lining. It adheres to the moist glue mixture. The container is gently rolled until the material covers the entire inside. Particles that do not adhere should be poured out and saved for later use.

A lid of clear acrylic, the same size as the round bottom, is routed to create a lip that will fit into the top of the container.

The edges of the lid are scraped, sanded, and polished.

The finished container with an unadorned lid.

Right: Tube containers can also be decorated by painting, engraving, and etching, and their lids may be varied too. Acrylic circles were adhered to this clear acrylic disc. The adhesive, PS-30, a two-part system, was mixed immediately before application. The syruplike glue produces a strong, clear bond. PS-30 was applied to the bottom side of each circle with a wooden applicator stick, and each circle was then pressed firmly onto the top of the lid. The same container—with a new top.

Many different types of containers are possible with acrylic tubing. Varying lengths of thinner diameter tubing are adhered to a piece of purple acrylic with solvent cement to create a pencil holder (*right*).

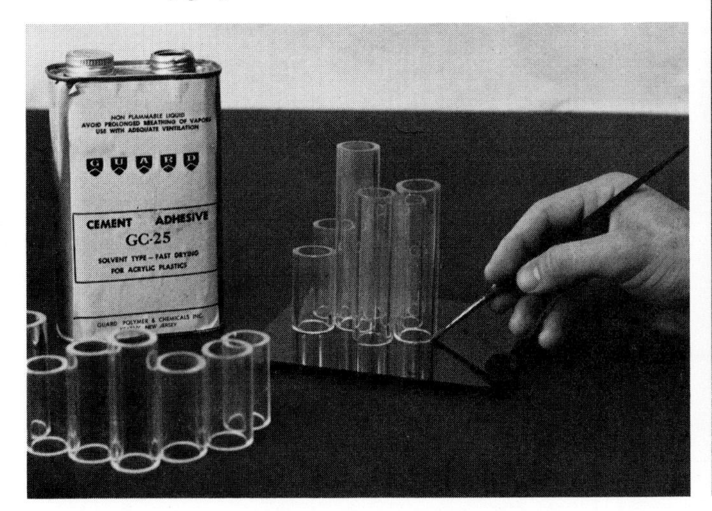

CONSTRUCTED ACRYLIC ORGANIZER

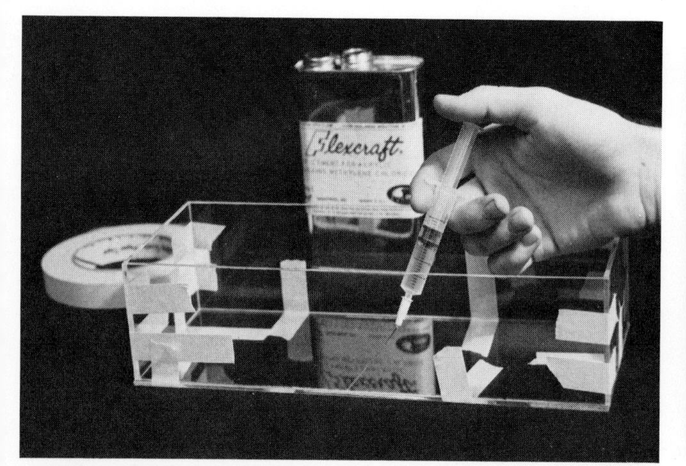

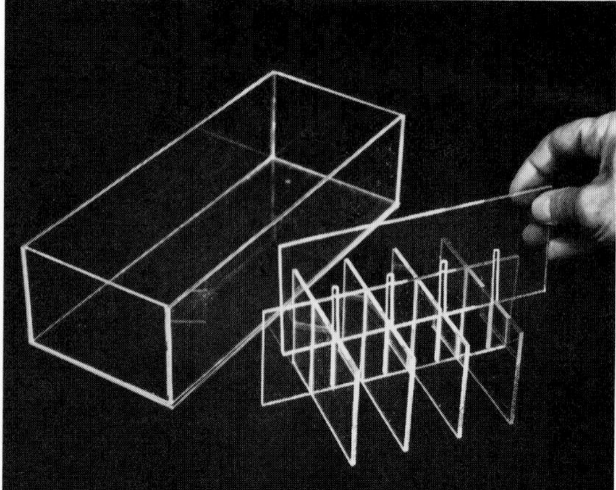

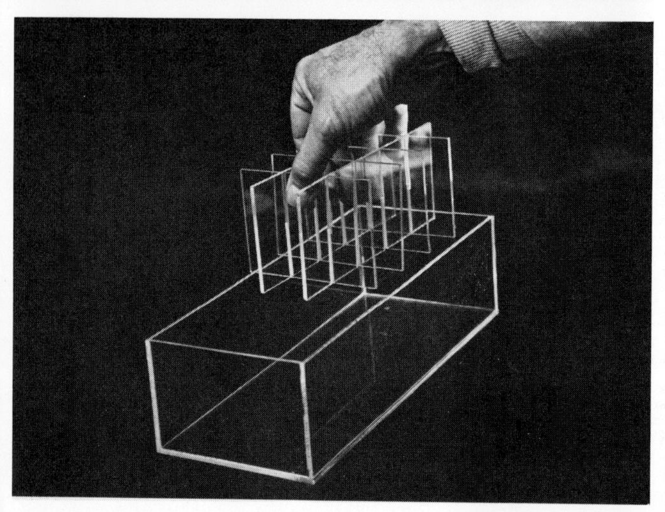

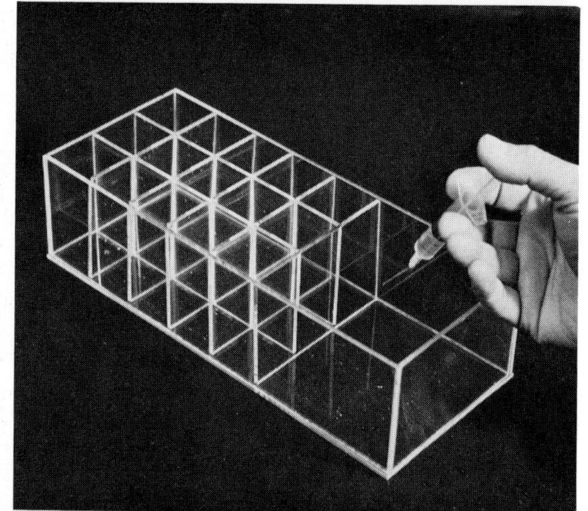

Top, left: The sides and bottom of this acrylic container were cut, scraped, sanded, and polished where they would be exposed. The pieces were then joined temporarily with masking tape, and solvent cement is being introduced into a joint through the flexible nozzle of a plastic syringe.

Top, right: The "organizer" part of this container consists of four pieces of acrylic that interlock through slits cut in them. These pieces may be glued together or allowed to remain unattached.

Center, left: Even if they are not glued, this section will stay in place. It will be held by a crosspiece of acrylic.

Center, right: The acrylic crosspiece is glued into place so that it braces the interlocking parts.

Right: Relief from clutter!

DECORATING A STYRENE CONTAINER

Plastic containers—like this one of molded styrene—may be decorated with other plastic parts. Plastic sign letters are arranged . . .

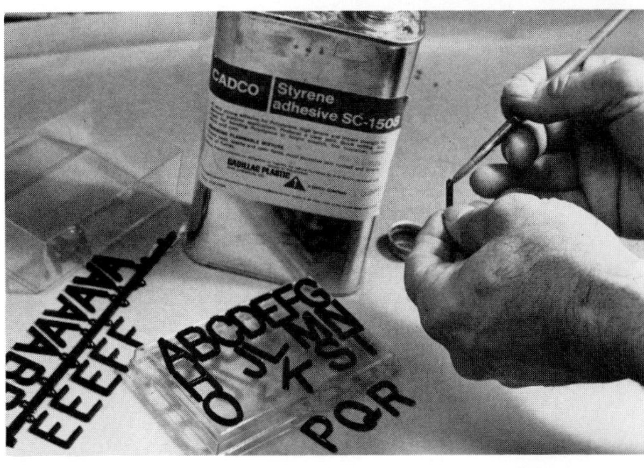

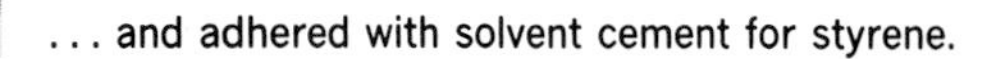
. . . and adhered with solvent cement for styrene.

The completed alphabox.

HEAT FORMING AN ACRYLIC BOWL

Acrylic should be allowed to remain in the protective paper that covers it as long as possible. The plastic can be cut very easily while the paper remains, and it is even easier to mark the protective paper for cutting and measuring than it is to mark the acrylic. Four slits are made at 45° angles to the four corners of this square piece of acrylic.

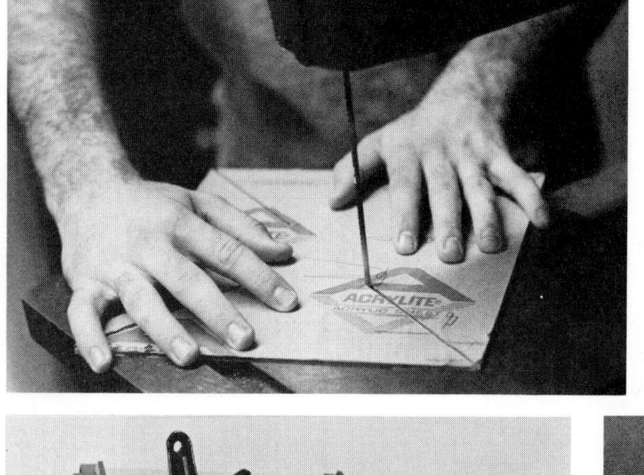

The acrylic is then peeled of its wrapper and placed in an oven that has been preheated to 275°F.

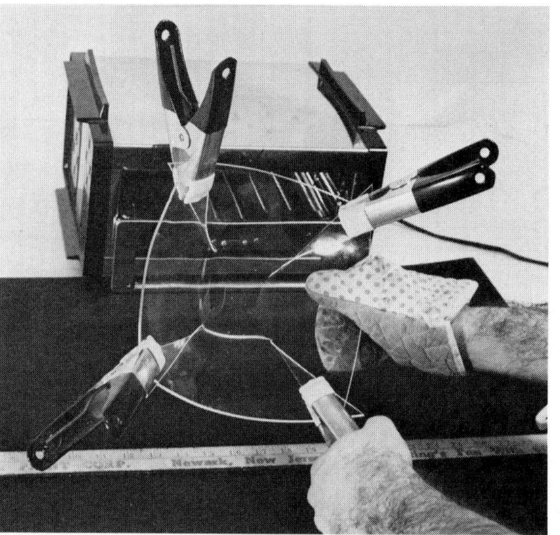

When the plastic turns rubbery, it is removed from the oven and shaped as desired. Here the sections at each corner are overlapped and clipped at the proper position. When cool, acrylic assumes its new shape. Because this is a thermoplastic, the material can still be returned to its original shape by reheating.

Right, center: Berry nice.

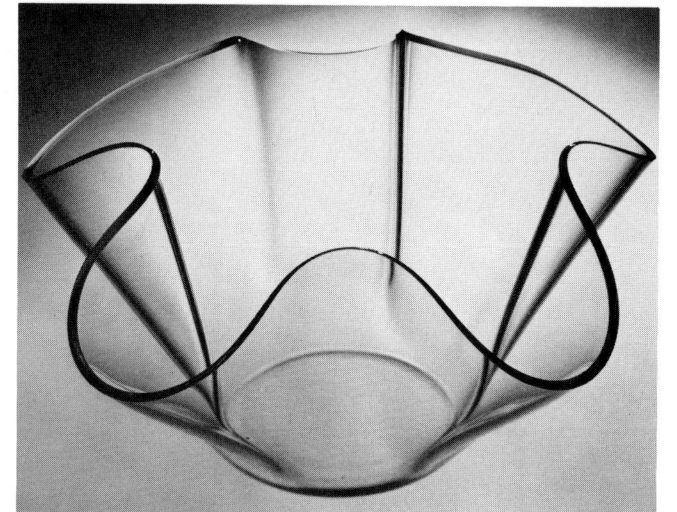

Right: A heat-formed acrylic "Handkerchief" bowl. *Courtesy Grainware Company*

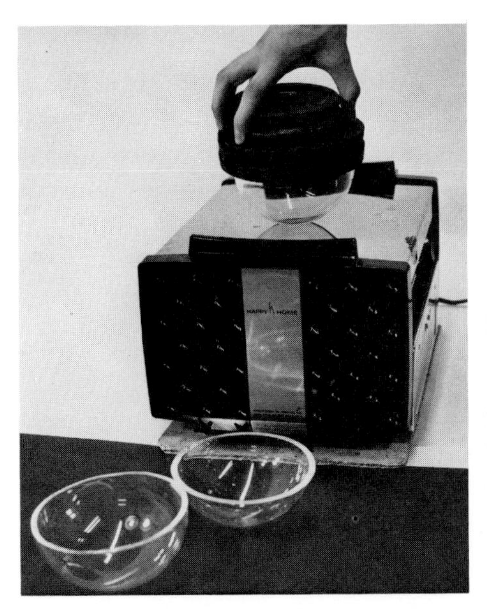

ETCHING ACRYLIC

Left: Many preformed acrylic shapes are available commercially. The hemispheres in this photograph, for example, are sold in a wide range of sizes. Just like any acrylic, preformed shapes may be reformed. A flat bottom is being formed on this hemisphere by heating it on the top of a broiler oven under the pressure of weights.

Right, above: The area to be "etched" is delineated by masking tape.

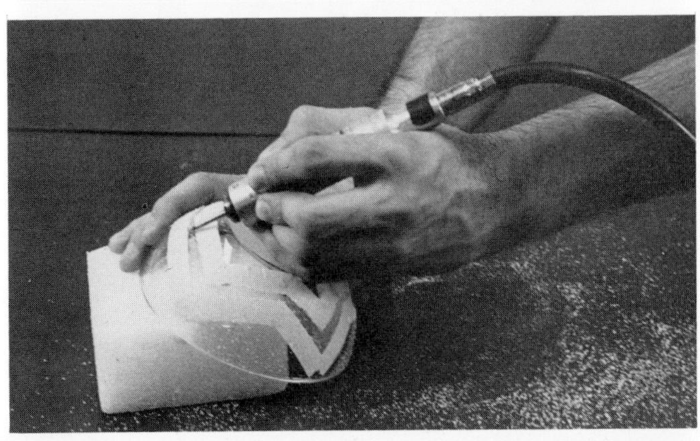

The acrylic is ground or etched with a grinding bit in a flexible shaft drill.

The finished bowls.

"Bottle with Stopper" (5" X 12"), by Daniella Kerner, consists of a vacuum-formed acrylic interior that has been copper electroformed and embedded in acrylic resin, which was then milled and carved. The stopper is cast acrylic, which has been carved and partially electroformed. (See chapter 8 on electroforming.)

Above: "Dream Container" (7½" X 7½" X 10"), by Daniella Kerner, is block cast acrylic, which has been machined and carved. The images are cast acrylic with vapor-deposited gold and silver. *Photos courtesy Daniella Kerner*

Procedures for Making an Acrylic Box

The steps in constructing an acrylic box are quite simple:

1. The acrylic is cut to shape on a saw
2. The edges are all dressed by scraping and sanding
3. Edges not to be glued—which will be apparent when the container is complete—may be polished to a high gloss
4. The pieces are aligned
5. Masking tape is used temporarily to hold all parts together for gluing
6. A solvent cement is applied along the edge of the joint with a syringe or brush
7. The cement is allowed to cure for at least one-half hour to assure a good bond

Liquid Plastics: Polyester and Epoxy Resins

Polyester resins are liquid plastics with a consistency like Karo syrup. They are available in a variety of formulations for as many different uses. For the container-maker, liquid plastics are particularly useful as casting materials. Although only polyester resins are expressly discussed here, epoxy resins are used in precisely the same way. The major differences between the two are that epoxies shrink less during the casting/curing process, and epoxies are more expensive.

Selecting a Resin

Because polyester resins are formulated for many different applications, it is important to select one that will be well suited for a particular use. In most container-casting operations, a *casting* resin will be the best type. Two good casting resins are Diamond Alkali's 6912 and Reichhold Chemical's Polylite 32–032. These resins are formulated to work well in castings less than three inches thick. Other resins are available for especially thin and thick forms.

Catalyzing

Polyester resin, unlike acrylic, is a thermosetting plastic. This means that once it is changed into a solid form, that will be its final shape—unless it is mechanically changed by machining or sanding. To convert polyester resin into a solid requires the addition of a catalyst. MEK peroxide (methyl ethyl ketone peroxide) is the catalyst most commonly used with polyesters; this triggers a chemical reaction in the resin that causes it to "cure" or harden. Catalysts must be added to all polyesters to induce the cure. And they must be added in the proper amount or the plastic will not cure properly. Most casting resins will respond to the tentative proportions set out here and calibrated to Polylite 32–032, but each company usually publishes suggested catalyses for castings of different thicknesses. The reason that any such measures must be called tentative is that successful castings depend upon the temperature of the room, the size of the casting (volume of the form), and the type of resin—not just the amount of resin and catalyst.

PROPORTIONS FOR MIXING POLYESTER RESINS

	POLYESTER RESIN (REICHHOLD'S POLYLITE 32–032 AT 70°F)	CATALYST (MEK PEROXIDE)
Thick castings (1/2"–3")	100 grams	1/4–1/2 gram

Thin sheets (40–250 mils; 10–60 sheets of paper)	100 grams	3/4–1 gram
Thin coatings	100 grams	2 grams
Thick castings (Over 3")	1 tablespoon	3 drops
	4 oz. (1/2 cup)	24 drops
	1 cup (1/2 pt.)	48 drops
Thin applications (castings and/or laminations)	2 tablespoons (1 oz.)	20 drops
	6 tablespoons (3 oz.)	1/4 teaspoon
	1 cup (8 oz.)	1/2 teaspoon
	1 pint (2 cups)	1 1/4 teaspoons

Additives

Colorants for use with plastic resins are available in many forms and types. There are powders, liquids, and pastes, with a color range from transparents and opaques to pearlescents and metallics. Unless the colorant is a powder, it should be one specifically formulated for use with polyester resin—and the same goes for epoxies. Colorants should always be added *before* the catalyst because the resin begins the hardening process once catalyzed, and adding the color first allows additional working time.

Other fillers impart different properties to resins. Cab-O-Sil and Aerosil make polyester thixotropic, or pastelike. When working on a vertical surface or applying resin in a sculptural effect, this pastelike form eliminates dripping. These fillers should also be added before catalyzing and should never make up more than 50 percent of the weight. Asbestos, which has been used as a filler in the past, should be avoided because of the recognized hazard to human skin and lung tissues. Calcium carbonate is a common type that extends the resin while making it more impact resistant.

Metal fillers most often come as powders. Again, they should be added before catalyzing in the following proportions: bronze filler—1 part to 6 to 7 parts resin; brass filler—1 part to 5 to 6 parts resin; copper filler—1 part to 4 to 5 parts resin; aluminum filler—4 parts to 5 parts resin; iron or lead filler—1 part to 7 parts resin. In addition to imparting a metallic color to resin, metal fillers also add some impact resistance.

Working Considerations

Plastic resins should only be used in clean, well-ventilated areas. All working surfaces must be clear, clean, and level, and it is wise to protect them with waxed paper, polyethylene, or Mylar. Room temperature should be between 70°F. and 75°F. Skin contact with all catalysts and resins should be avoided; wear disposable gloves. Mixing containers should also be disposable. Cans and liquid-proof paper cups are suitable for mixing operations. Resin and catalyst are more easily measured by weight, but they may also be measured by volume using measuring cups and spoons if no scale is available.

Molds for Resins

In container-making, molds for casting polyester resin are a great resource. Working from standard tubes, or sculpted originals, molds provide an opportunity to make multiple forms with ease. Given the variety of additives available for use with polyester resin, there is still enormous room for variation even with a single mold.

Molds may be rigid or flexible, temporary or permanent, of one piece or several. They may be made from wax, aluminum foil, plaster, plastic, glass, rubber, vinyl, or room temperature vulcanizing (RTV) silicone rubber. For most applications, this last material is the most versatile, and it is very easy to use because of its flexibility. Pieces with undercuts can be released without locking into the mold.

All molds begin with an original form. That piece is placed within a leakproof container of plastic or paper, which may be sprayed with a mold release agent to prevent sticking. Most mold materials must be mixed first; RTV silicone, for example, is a two-part system like polyester resin. The mold material is then poured around the original. Once the mold material cures, the outer container is removed; the original is released from the mold. It now is ready to accept catalyzed resin, which will harden to create a duplicate of the original form.

Making Resin Castings in Molds

Once the mold is complete, the casting process is really quite simple. Some mold materials require coating with a release agent—most often a silicone spray—which will prevent the resin from sticking to the mold surface when it hardens. Other materials—like RTV silicone rubber—do not need any treatment; they separate naturally. To make a casting, the polyester resin should be catalyzed considering the volume (mass) of the walls of the casting, and poured into the mold. Once the resin hardens, the form should be demolded and finished if necessary.

Finishing Polyester Forms

Polyester forms, like those of acrylic, may be readily cut, drilled, tapped, sanded, and polished. Carbide-tipped metalworking blades with uniform teeth work best and, as with acrylics, solid polyester forms should be fed through power machinery slowly to avoid excessive heat buildup.

When drilling and tapping, polyester should be lubricated with oil coolants or a solution of mild soapy water. Wet-sanding is recommended to avoid excessive polyester dust in the air. When sanding, it is best to progress from coarse grits, to remove deep marks, to the finer papers (i.e., 150, 220, 400).

The two-wheel buffing system, with loose 10-inch buffs rotating at 2,000 surface feet per minute, is excellent for finishing polyester to a high gloss too. But polyester buffs should never be used with acrylic. Polyester builds up on the wheels. The heat created by friction transfers the polyester to the acrylic and discolors the acrylic. Also, never buff polyester when it is not completely cured. This really gums the wheel and ruins it.

Polyester may also be brought to a high "polish" by painting the surface with very highly catalyzed resin.

Procedures for Casting a Polyester Container

The steps in constructing a cast polyester container:

1. An original is constructed or sculpted
2. A liquid-tight container is made to house the original and mold material
3. The original is placed in the mold container
4. The mold material is mixed according to manufacturer's instructions
5. The mold material is poured into the container around the original and allowed to cure
6. When the mold has hardened, the outer container is removed and discarded if disposable, and the original is demolded
7. The mold is prepared with release agents, if necessary
8. Polyester resin is prepared in a disposable mixing container by adding proper fillers and colorants and mixing them thoroughly
9. Catalyst is added according to the thickness of the form
10. Catalyst and polyester resin are mixed thoroughly to assure complete distribution
11. The catalyzed mixture is poured into the mold
12. Once the polyester has cured completely, the solid form is demolded (this will often be the last step)
13. If necessary or desired, the molded container may be machined and/or finished

CAST POLYESTER CONTAINERS AND LIDS
by Ralph and Sylvia Massey

The first step in casting any form is to make or obtain a mold. Ralph Massey is making a mold for a cylindrical container from a section of acrylic tubing. This section is closed at one end by a round piece of polished acrylic—just like the cylindrical container constructed above. A wooden center plug is at left. The pieces of wood at the top of the center plug will be used to suspend the plug in the mold. The plug takes up space that would otherwise have been filled by RTV silicone mold material. Not only is this expensive material saved, but demolding is made easier as well. At the right is the mold housing.

The container has been placed bottom down inside the mold housing. (The masking tape at the corners of the mold housing prevents leakage of mold material.) The center plug has been suspended inside the mold housing too, and its support blocks can be seen resting on the top of the housing. Catalyzed RTV silicone is then poured into the mold, around the acrylic tube.

After the RTV silicone cures, the mold housing is removed, and the center core is removed, and the original pattern—the acrylic cylinder—is also removed. This shows the center core end of the mold. The mold housing is retained and will be used to support the mold during each casting.

Ralph Massey demolds a polyester resin casting. This is the other end of the mold.

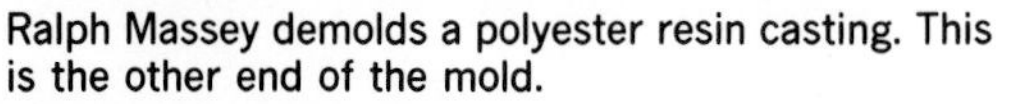

The container lids are first modeled in clay, and then molds are made in a fashion similar to that shown above. Because of the complexity of the forms, however, a two-piece mold is necessary. The mold is cut apart after the RTV silicone cures. Note the notches at the edges of the mold. These allow the halves to be aligned perfectly to create a perfect casting, rather than a piece with half askew.

The castings are finished with file, sandpaper, and steel wool.

Ralph Massey's lecherous egg—"Bouquet"—and companion were painted with acrylic colors.

Sylvia Massey carefully sculpts her animal forms in clay—with the aid of a magnifying glass and a high intensity lamp.

This casting also requires a two-part mold. The boar is still waiting to be demolded. Note the conical hole at the bottom of the picture; this is really the top of the mold, and the hole is where resin enters the mold.

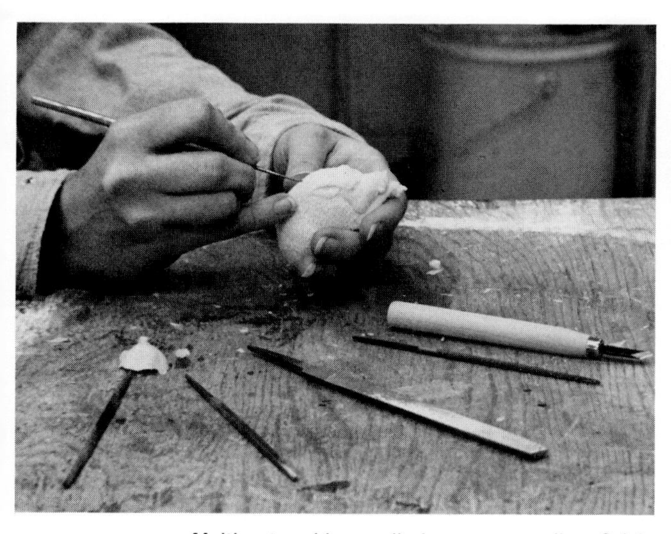

Multipart molds usually leave a seam line. Sylvia Massey scrapes away all evidence of the seam, then files and sands the tiny beast.

The "Sleeping Boar" rests on a nest on top of the cast cylinder.

It is then colored in lifelike shades of brown acrylic paint.

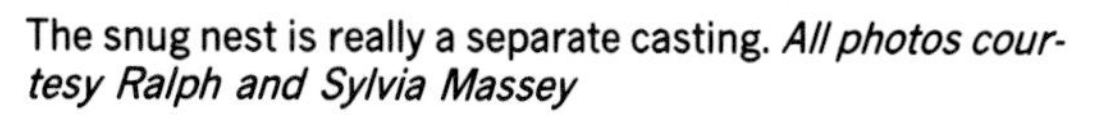

The snug nest is really a separate casting. *All photos courtesy Ralph and Sylvia Massey*

Commercial cast containers. *Courtesy Crayonne*

Polyester and Fiberglass

Fiberglass-reinforced polyester is another structural material easily adapted to container-making. The basic process is quite simple. Fiberglass strips, or pieces cut from fiberglass sheet, are dipped into catalyzed and colored polyester resin. These saturated pieces are then draped over or into molds. The impregnated fiberglass will cure in the shape of the mold. Many types of forms may be used as molds; ceramic or glass bowls are particularly useful. Simply cover the mold form first with plastic wrap to protect it from the polyester resin (plastic wrap also acts as a release agent). Layer the strips of polyester-saturated fiberglass over the surface.

Once the material has cured thoroughly, the piece may be machined just as could any polyester form. Edges and surfaces may be trimmed, scraped, sanded, and polished.

POLYESTER RESIN AND FIBERGLASS CONTAINER

The combination of polyester resin and fiberglass results in a very strong structural material—and the process of fiberglass reinforced polyester forms is very simple. The first step is to prepare the mold form. Use a form with no undercuts, but first cover it with plastic wrap, which will act as a separating agent and prevent the resin from sticking to the model. Strips of fiberglass should then be cut from the larger sheet. The resin may be colored, if desired, and catalyzed. In case of spills, acetone is the solvent for polyester resin.

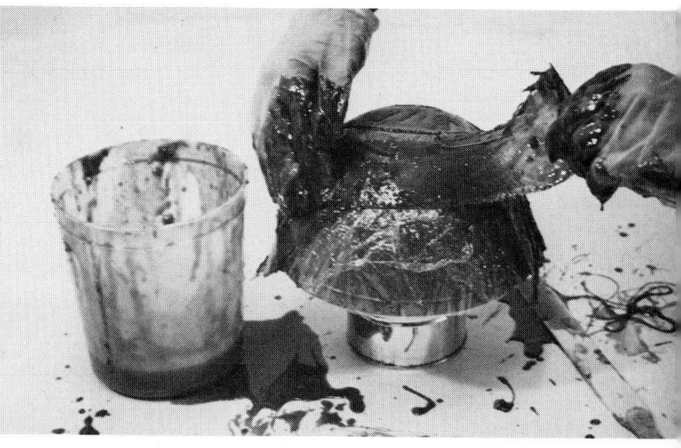

The fiberglass is dipped into the catalyzed resin, until saturated, and wrapped around the bowl. Apply one layer in one direction, and the next layer perpendicular to the one before it, and so on.

Once the resin has cured, the form may be machined just like wood or metal. The edges may be trimmed on a band saw, and . . .

. . . surfaces may be sanded by machine or by hand.

To remove scratches created by sanding and machining, buff the form to a high gloss on a muslin wheel with rouge. Never use the same wheel for acrylic and polyester since gummed surfaces will result.

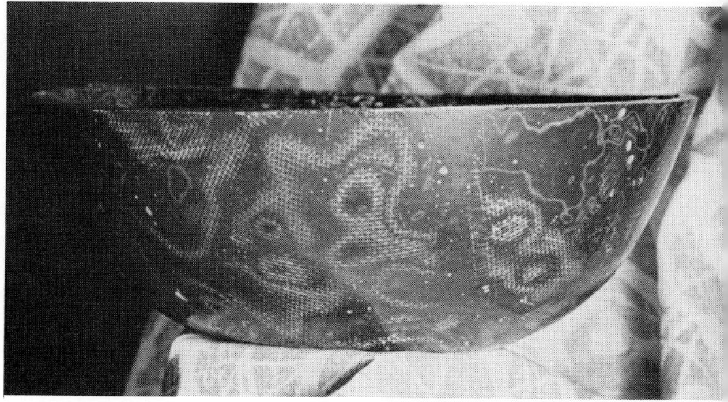

The finished container.

Other Plastic Materials for Container-Making

Fusible Plastic Pellets and Tiles

Fusible plastic pellets and tiles—like Dec-Ets and Poly-Mosaic tiles—also extend the range of container-making possibilities. Both pellets and tiles (which are 3/4" square) are available in a complete range of colors. Both are readily fused in a home oven at temperatures below 350°F. These materials may be placed on oven trays or aluminum cookie sheets for fusing. After the tiles have fused together, the form may be lifted, while hot and rubbery, from the tray and pressed into or over a mold. The hot flexible forms may also be held in place and rapidly cooled by dipping in warm water.

When completely cool, the fused and formed shell may be sanded and machined. Polishing can be accomplished by gently passing the areas to be treated through the flame of a propane torch.

The result is a richly colored transparent/translucent form, like stained glass in its brilliance, but much easier to create.

A CONTAINER OF FUSIBLE PLASTIC TILES

Fusible plastic tiles, Poly-Mosaics (Poly-Dec Co., P.O. Box 541, Bayonne, New Jersey 07002), are available in many brilliant transparent colors. Each tile is ¾" square, and may be fused in a home oven at 350°F., on an aluminum cookie sheet or glass oven tray. Poly-Mosaics may be arranged in any manner desired, and tiles of different colors (or the same color) may be overlapped to create thicker forms or to achieve depth. Traditional tile cutters are an effective cutting tool.

After arranging the tiles on a cookie tray, they may be placed in the oven. The degree of fusion is determined by the length of time the tiles remain in the oven. Variations in texture may be achieved by varying the length of firing.

After these tiles were partially fused, their tray was removed from the oven for a few moments. Edges of the form were shaped with a metal spatula to create a perfectly round form, and the piece was then returned to the oven for a few moments. It was then removed again, and the piece was separated from the cookie sheet with the flat blade of an aluminum spatula and the still hot tile form was laid over a ceramic bowl, which was placed on the tray and then placed in the oven with the tile form on top.

As the Poly-Mosaics melted further, their shape conformed to that of the bowl, resulting in a colorful and easily constructed fused plastic bowl.

Polystyrene and Polyurethane Foams

Both polyurethane and polystyrene foams should be considered as elements in the media arsenal, too. Both foams are available in many densities. The harder ones are still easily carved, and once shaped they readily accept a wide variety of surface treatments. Acrylic gesso and modeling paste may be used to fill the pores and finish the surfaces, which then might be decorated with acrylic paints. Polyurethane may be covered with polyester-impregnated fiberglass. And although polystyrene will dissolve under polyester resin, epoxy is completely compatible with this foam. In another type of combination with fiberglass, the carved foam might serve as a mold which fiberglass can be draped over, into, or around.

8 METAL CONTAINERS

Contemporary craftsmen benefit from sophisticated techniques and advances in metallurgy that their ancient predecessors pioneered. While the range and variety of materials and processes have been expanded, the essentials remain unchanged.

An extensive assortment of materials confronts us today. Brass, copper, silver, gold (in a dozen colors), platinum, steel, iron, and bronze are widely available. Most are marketed as wire, sheet, precut shapes, blocks, and castings. In addition, each category embodies a further range of specialty products.

Yet, the working processes such as casting, raising, repoussé, fabrication, and finishing are age-old. Though the observation is banal and repetitive, that is only because it is true. We have added refinements and a few laborsaving devices, but basic techniques are still basic, and we have probably lost more old wisdom than new technology has provided.

Casting

For over four thousand years, craftsmen have used the lost-wax casting process. The technique appears to have developed simultaneously in many parts of the world. It was known in ancient Greece, Egypt, Italy, South America, and Africa. Artisans throughout the world still use it today.

The process, essentially, involves an original of wax, acrylic, paper, cloth, or other organic material, which is invested or contained in an investment or shell. Most often the investment is a combination of clay, charcoal, and plaster. The entire form is then heated so that the original will burn out, leaving a cavity within the investment. Molten metal is poured into that cavity and, when the metal cools enough to lose its cherry color, the investment is placed in water. This cools the piece quickly and causes the still hot investment to crack apart to reveal the casting. When completely cool, the cast form may be finished by sanding, filing, buffing, plating, and coloring as desired.

This technique does have the disadvantage of destroying the mold, but there is a decided advantage to being able to work the original in wax or another malleable material and preserve fine detail permanently in cast metal. One wax formula used today: 100 parts white beeswax, 100 parts paraffin, 50 parts petroleum oil, 50 parts petroleum jelly, and 10 parts lanolin is very readily worked and is especially responsive and true to the artist's idea. Of course, once the metal form is complete, it can, in turn, be replicated by making a mold of it, or it may be allowed to remain a unique piece.

Procedures for Lost Wax Casting

The basic steps in lost-wax casting are easy to follow:

1. Make an original in wax, acrylic, papier-mâché, cloth, or any other organic material (though wax is probably best)
2. Attach sprues, which form canals to the surface of the investment and allow for the exit of the original material and entry of the metal
3. Coat the original with a solution that inhibits bubbles from forming
4. Choose a flask in which to place the form and pour the investment
5. Place the original in the flask, add the investment, and tap the flask to dislodge any trapped air
6. Allow the investment to harden
7. Place the flask in an oven or kiln, sprue holes down, to melt or burn out the original
8. When the original has been completely burned out, remove the flask from the oven or kiln
9. Melt the casting metal in a fluxed crucible and inject it into the cavity, either by gravity, or according to the operations of a centrifugal caster

CAST IRON CONTAINERS by David Luck

David Luck's cylindrical cast iron containers are manufactured by a negative molding process. In this process, no pattern is made. Instead, the mold for the molten iron is carved directly into slabs of self-setting core sand. This photograph shows sand slabs ready for carving. One slab has had its center drilled out with a core drill made from a large piece of copper tubing at left. The abrasive stones and sanding discs on the right will be used to further shape the core sand. As the slabs are carved, they will be stacked over the preformed core to create a hollow mold area to accept the molten iron. Different kinds of sand are used to create different textural effects in the iron.

The mold is being built from carved sand slabs, which are stacked around the preformed core.

After carving of the sand slabs has been completed, they are pasted together over the core. The end slabs have sprues cut into them and pouring cups are pasted on. The finished molds are bound together using iron binding wire to withstand the internal pressure of the molten iron.

The finished molds for the iron are rammed in green sand (damp) by students of the University of Iowa Graduate Sculpture Department. Professor Julius Schmidt is the head of this department.

A sculpture student charges scrap iron into the top of the cupola furnace. (Please note that he is using the trash container only as a container to hold the scrap iron pieces.) The cupola furnace is a refractory lined vertical steel cylinder with a well at the bottom where the molten iron can collect. Tuyeres along the middle section allow air to be introduced into the furnace. The high heat needed to melt the iron pieces (fed in from the top) is created by the reaction of coke fuel (fed in from the top as well) and blower-forced air. Holes are provided in the well section (bottom) to allow slag runout and to tap the molten iron.

Another sculpture student taps out the molten iron into a ladle by driving a steel stake through the clay-plugged tap-out hole at the bottom of the well, allowing the iron to gush out.

The molten iron is poured from the ladle into the waiting molds.

After the iron cools, the mold sand is broken away, exposing the rough casting. The sprues are cut off and the container is finished by grinding, polishing, and chasing.

A cast iron container by David Luck, made by negative carving in sand block and casting around a self-setting core. This photograph sequence was taken by David Luck.

CAST ALUMINUM CONTAINERS by Don Drumm

Sand casting is another expendable mold technique. The first step is to make a model of the form to be produced—this is the pattern for the mold. Don Drumm makes his pattern by hand, often in clay.

It is then placed into a metal mold frame, where sand is sifted around the form and tamped down firmly. Many substances can be used to make the mold besides sand. They include cement, fireclay, and plaster. All must be bonded in some way so that the mold material has the strength and dimensional accuracy to retain the shape of the pattern. Bonding may be achieved chemically or by drying the material. In green (damp) sand molding, drying affects the bond.

After sand has been sifted around the pattern, it is compacted by pressure or vibration, and dried. The mold is then opened, and the pattern is removed. Then the mold is closed, leaving a cavity into which metal is poured. Where necessary, a core may be inserted to achieve a space.

The crucible furnace is then charged with metal, and a ladle is used to introduce the molten aluminum into the mold.

The forms are finished by filing and grinding. Where necessary, pieces are arc-welded or heli-arc group-welded by an industrial process using inert gas.

Don Drumm's sand-cast aluminum casserole is food-safe and great for cooking because the ¼" walls distribute heat evenly.

A cast pewter box by Patricia J. Daunis-Dunning. *Courtesy P.J. Daunis-Dunning*

Above: Cast aluminum muffin tray by Don Drumm.

Left: Both halves of Don Drumm's large deep cooking pot can be used for cooking (9" X 9").
Photos courtesy Don Drumm

Bottom, left: "Touché," cast pewter, brass hinges, and abalone fingernails, by Paul A. Diekmeyer. The top of the hand and the small finger of this container open as a unit. The piece is on permanent exhibition at the Oregon Museum of Art. *Courtesy Paul A. Diekmeyer*

Below: "Two for the Road" (3" X 1½" X ⅞"), a cast brass and silver box by Fred J. Woell. *Courtesy Fred J. Woell*

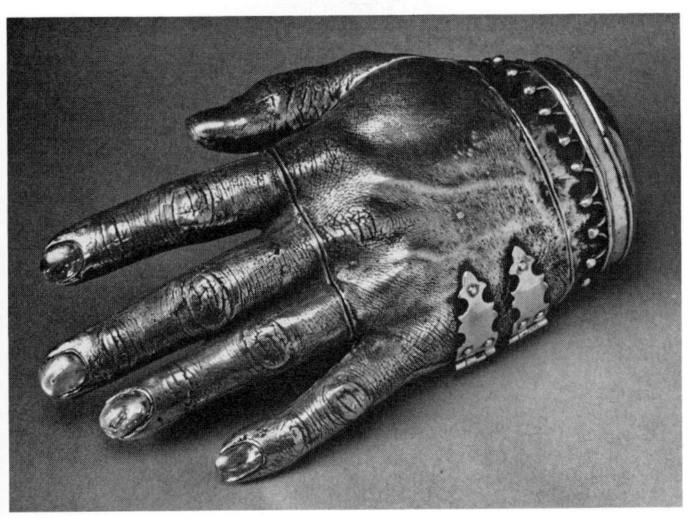

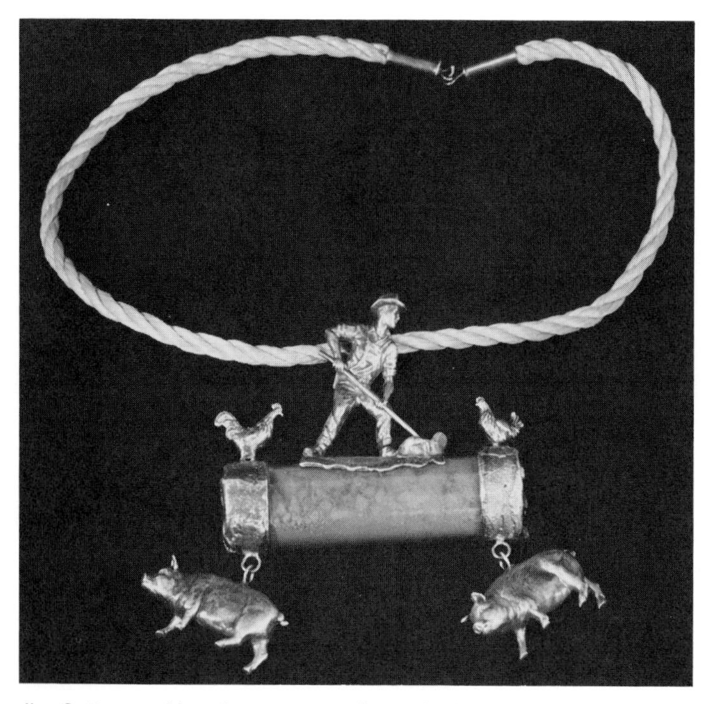

Jim Cotter combines lost-wax cast silver, plastic, dirt, and rope in "The Great American Farmer" (5" X 6"). *Photo by Tom Lamb*

"Container Ring No. 1" (1973), by Harlan Butt. Cast silver inlaid with ivory and shell. *Courtesy Harlan Butt*

Three boxes (3" high), of cast sterling silver, 14-karat gold, and bloodstone, by Earl Krentzin. *Courtesy Earl Krentzin*

Forming and Repoussé

The most basic structural and decorative metal processes are forming and repoussé. Forming is probably best known as raising. But before looking into these processes, it is best to say a few things about the physical properties of metals that are most often formed: copper, brass, bronze, silver, and gold.

These metals are relatively soft; they can be hammered and worked easily. However, hammering transforms and distorts the crystalline structure of the metal. It becomes flatter, harder, more brittle. Of course, that may be part of the goal. Hammering shapes metal. Striking one area more than another, in addition to modifying the elemental structure, will create a thinner sheet or perhaps a greater curve. Eventually, however, the metal will become brittle and it will crack. To avoid this, and to allow the metal to be worked most fully, it must, in a sense, be "released" periodically. This process is known as annealing; the metal is heated to a dull red and allowed to cool. Heating allows the molecules to become reoriented, and a new crystalline structure results with the metal becoming soft and malleable once again. This hammering and annealing process must be repeated often. When nearly completed, the piece is hammered and not annealed so that it remains hard and will not bend easily.

Hammering techniques can produce a variety of shapes. And there are an assortment of hammering styles. *Sinking, blocking,* or *hollowing* is accomplished by hammering a piece of metal on the inside of the proposed container, forming it to conform to the hollow of shaped anvils made of wood or metal. *Raising* is the opposite; hammering is done on the outside of the metal while it is held over a stake or anvil. The hammers, once made of smooth, polished stones, are now highly polished metal and are available in a wide range of sizes, shapes, and weights.

Raising

To raise a flat piece of metal into a shape, metal thickness should begin at 14–18 gauge. To determine the size of the initial piece of metal, add the desired depth to the intended diameter to determine the raw width. The form is raised by holding the metal in one hand, with the metal resting on an anvil, and the hammer in the other hand, ready to strike. Every time the metal is struck, the piece is rotated. Each blow should be administered at the same angle.

When the metal begins to resist hammer blows, it requires annealing—copper at 700–1200°F., bronze at 800–1100°F., sterling silver at 1200°F. —or until the metal is a dark red.

To anneal, the form is placed in an annealing pan filled with lumps of pumice, or a refractory substance capable of retaining and distributing heat without melting. The metal is usually heated with a torch (propane will do) in a slightly darkened area so that the color can be judged. The flame is moved slowly over the entire form so that the piece is heated evenly. To focus too long on any one spot might warp or melt the metal. (In processes where only one spot is worked at a time, however, it is possible to heat each area individually.)

The piece is quenched immediately after annealing in a pickling solution, usually a dilute acid or other substance that cleans the metal by removing oxides that build up when heat is applied. A typical pickling solution is ten parts water into which one part sulphuric acid is poured slowly. (Never the other way around because of the splattering that may result.) There are also several effective acid substitute pickling solutions that are safer, although slightly slower working. Pickle should be contained in glass, polyethylene, or enamel, and it can be stored and reused. After removal from the solution, the object should always be rinsed thoroughly.

After annealing, the raising procedure continues until the form takes its shape, or until further annealing is required.

A heavy rawhide hammer and a mushroom-shaped anvil are used to smooth blow marks in the final stage of raising.

MARY KRETSINGER'S FORMED AND CONSTRUCTED CONTAINERS

Mary Kretsinger begins by cutting a predetermined shape out of a sheet of brass; this will become the base of the container. She then roughs out the shape with a ball peen hammer on a wood or lead block.

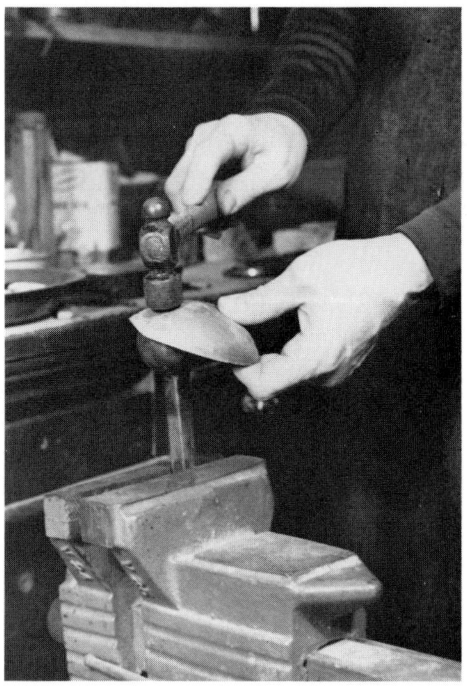

The form is further refined by raising over a steel stake made from a ball bearing. The brass must be annealed periodically to soften it and allow for further working.

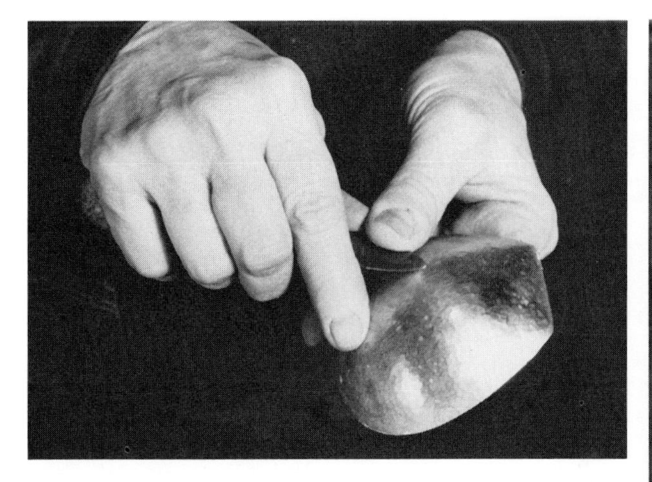

After raising, the form is pickled and rubbed down with #0000 steel wool. Scratches and rough areas are smoothed with a burnisher, and the form is rubbed with steel wool again.

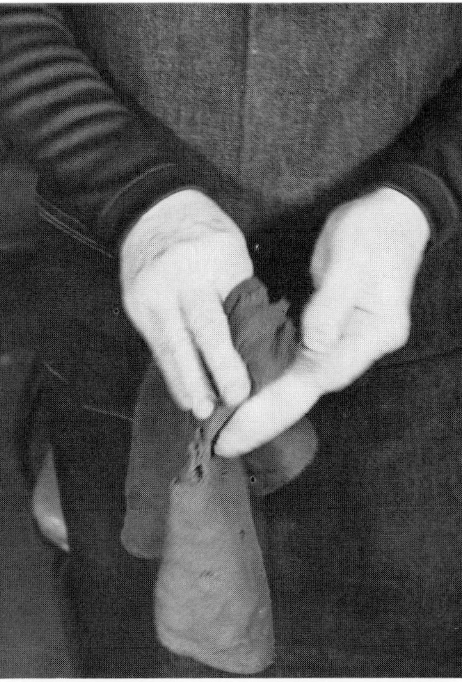

Mary Kretsinger then rubs the brass with a half-and-half mixture of jeweler's rouge and superfine pumice on a soft rag.

She then cuts a piece of brass slightly larger than the shaped form, cuts out its center, and prepares the piece for soldering. This will become the lip or rim of the container's bottom half. Here she is burnishing the surface of the brass so that it lies perfectly flat.

She carefully matches both bottom and rim so that there are no gaps, and fluxes the surface in preparation for soldering. The flux may be either a half-and-half mixture of boric acid and methyl alcohol, or Battern's self-pickling flux.

Snippets of Baker 569A 10k soft gold solder are then applied around the entire joint. Mary Kretsinger uses gold solder and fills in any gaps between rim and bottom with it.

The form is banked with refractory material and heat is applied to the form, causing the solder to flow into the joint.

After soldering, the piece is pickled and cleaned in Sparex®. This view shows the inside (concave) surface of the bottom, and the rim that will meet the lid.

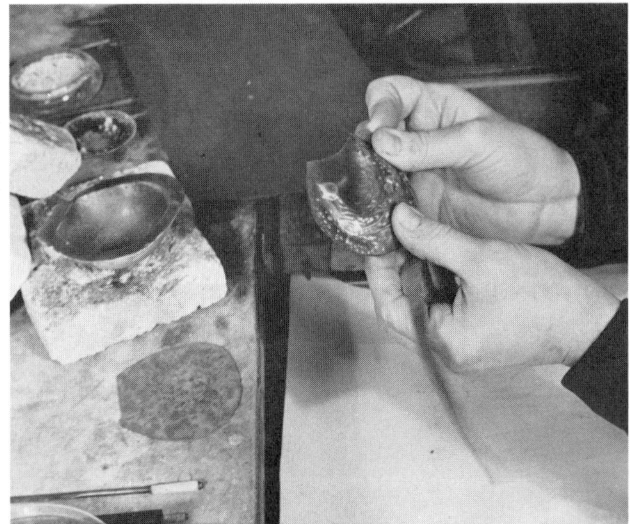

The lid of this container consists of a raised, enameled form attached to a flat piece of brass by a brass bezel. The bezel is a metal rim soldered to the flat piece and bent over the enameled form to hold it in place. Here Mary Kretsinger wraps copper foil around the edge of the enameled form to determine the length of the bezel.

She then measures the length on a sheet of 20-gauge brass, which has been rolled down to 22 gauge . . .

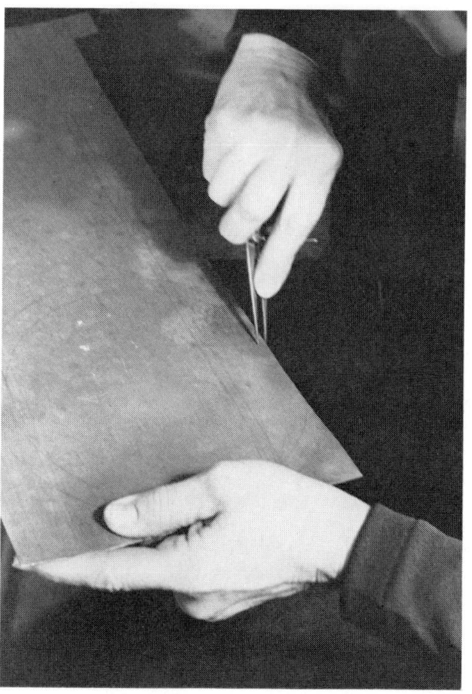

. . . and scribes a line ¼" from the edge with dividers.

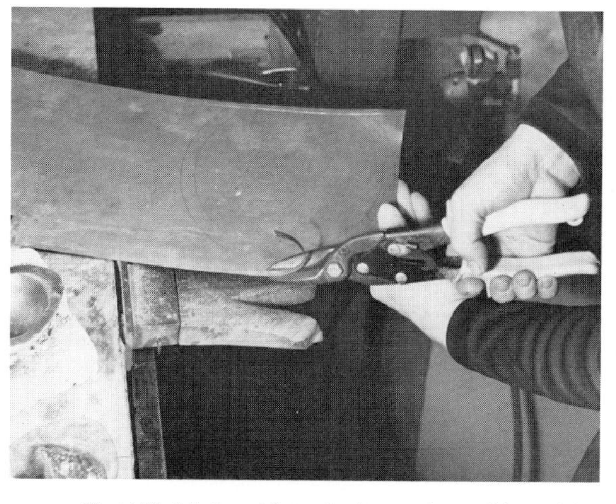

That ¼" strip is cut from the brass sheet with straight-cut aviation shears.

She carefully and painstakingly bends the bezel to shape and fits it to the enamel form and brass lid. This stage is crucial, because warpage will prevent a good fit. Mary Kretsinger uses pliers with smooth jaws to avoid scratching the bezel.

The bezel (alongside the rim plate here) is burnished to make it completely flat, and its ends are soldered together.

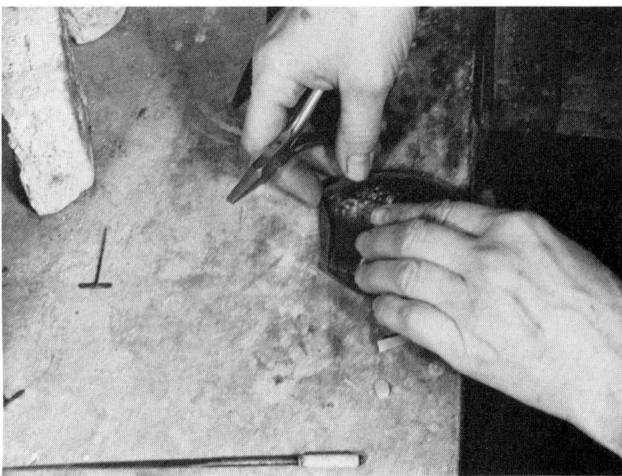

The bezel is then cleaned, set in place, fluxed, and soldered to the flat top lid. After soldering and cleaning, the enameled form is set within the bezel, and the bezel's sides are burnished down so that they hold the enameled form securely.

Top and bottom halves of the container are attached with a seven-piece hinge made from copper tubing which has been cut, pinned, and soldered into place. The box top is cloisonné enameling by Mary Kretsinger.

The inside of another container by Mary Kretsinger.

Repoussé

Repoussé is a form of surface decoration that may be combined with raised forms, or flat pieces of metal that are soldered together (fabrication). In repoussé, the design stands out of the surface in relief. The process begins with a sketch scratched onto the metal with a sharp, pointed instrument called a scriber. The areas to be raised are worked with tools called modeling cushions or bossing tools, which are pressed into the metal with hammer blows. During this process, the metal rests in pitch, which provides just enough resistance to allow the metal to be formed while retaining firm control of the form and the depth of relief. Alternative supports are sandbags or leather pads.

A small, chisel-shaped tool, called a tracer (also powered by hammer blows), is used to define lines and areas around sections in higher relief. Chasing, or hammer and punch work on the surface, further refines these shapes.

To prepare pitch, heat it in a double boiler and then pour it into a hollow form. After it has been poured into the tray or whatever will be used to contain it, the slightly oiled metal can be worked directly on the pitch. Oiling aids later removal.

Repoussé tools are not cutting tools like those used for engraving. They are shaping instruments, and they are available in over fifty shapes for as many functions. Scribers, tracers, bossing tools, and modeling tools are the most common; others have been developed for specific applications. All are used with repoussé or chasing hammers that have steel heads and polished faces.

To work the metal, the form is first placed into the softened pitch deeply enough so that all surfaces to be worked are in contact with the pitch. The design is scribed onto the surface and then traced with the tracer. The tracer should be held in the left hand between thumb and index finger, using the third and fourth fingers as guides, levelers, or balancers while resting on the metal. The tracer is held nearly perpendicular to the metal surface, slightly back from the direction it is to move in, and propelled by blows from a hammer held in the other hand. With each blow the tool should move slowly, continuously, along the line to be scribed. The result should be an unbroken line, not a jagged one. Once this outline has been established, modeling tools are used to emboss or repoussé the enclosed surfaces.

When work has been completed on one side, the metal is reversed. To release the metal, the pitch is heated with a torch. Hot paraffin, benzine, or turpentine on a cloth removes all traces of pitch from the metal.

If the piece requires annealing (when the metal begins to resist hammer blows), this should be done before reversing the piece for further repoussé on the other side. The reversing process may have to be done many times, but it results in a much more consistent surface.

Chasing is continued on the front surface until the desired degree of relief has been achieved. When working, always try to work the entire piece a little at a time, rather than one part all at once.

Richard R. Dehr set this raised silver form into pitch to begin repoussé. *Courtesy Richard R. Dehr*

A constructed silver container with repoussé designs, by Richard R. Dehr. *Courtesy Richard R. Dehr*

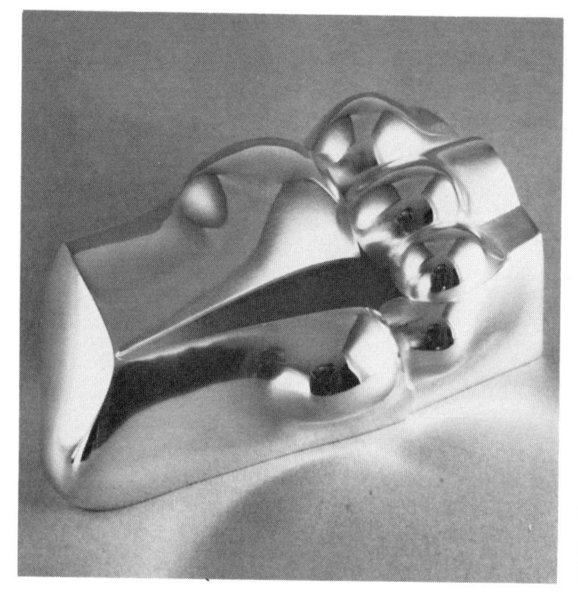

"Slit Form" (1973) (2½" X 5"), of formed and constructed 18-gauge sterling, by Helen Shirk. *Courtesy Helen Shirk*

"Copper Container I" (1974) (7" X 6"), formed and constructed of 18-gauge and 14-gauge copper with steel wool, oxidized finish, by Helen Shirk. *Courtesy Helen Shirk*

Fred Fenster's raised and cast "Lip Pot" (7" high).

"Ikon" (4¾" high), formed and constructed sterling, 14-karat gold, and Mexican opals, by George P. van Duinwyk. *Courtesy G. P. van Duinwyk*

A raised pewter vase (9" high), by Fred Fenster. *Photos courtesy Fred Fenster*

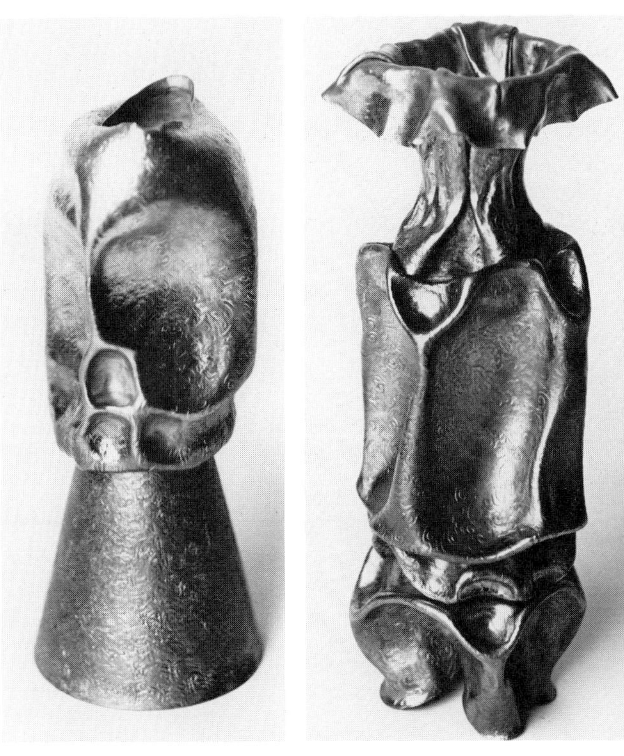

Two pots of raised, shaped, crimped copper (12" and 14" high) by L. Brent Kington. *Courtesy L. Brent Kington*

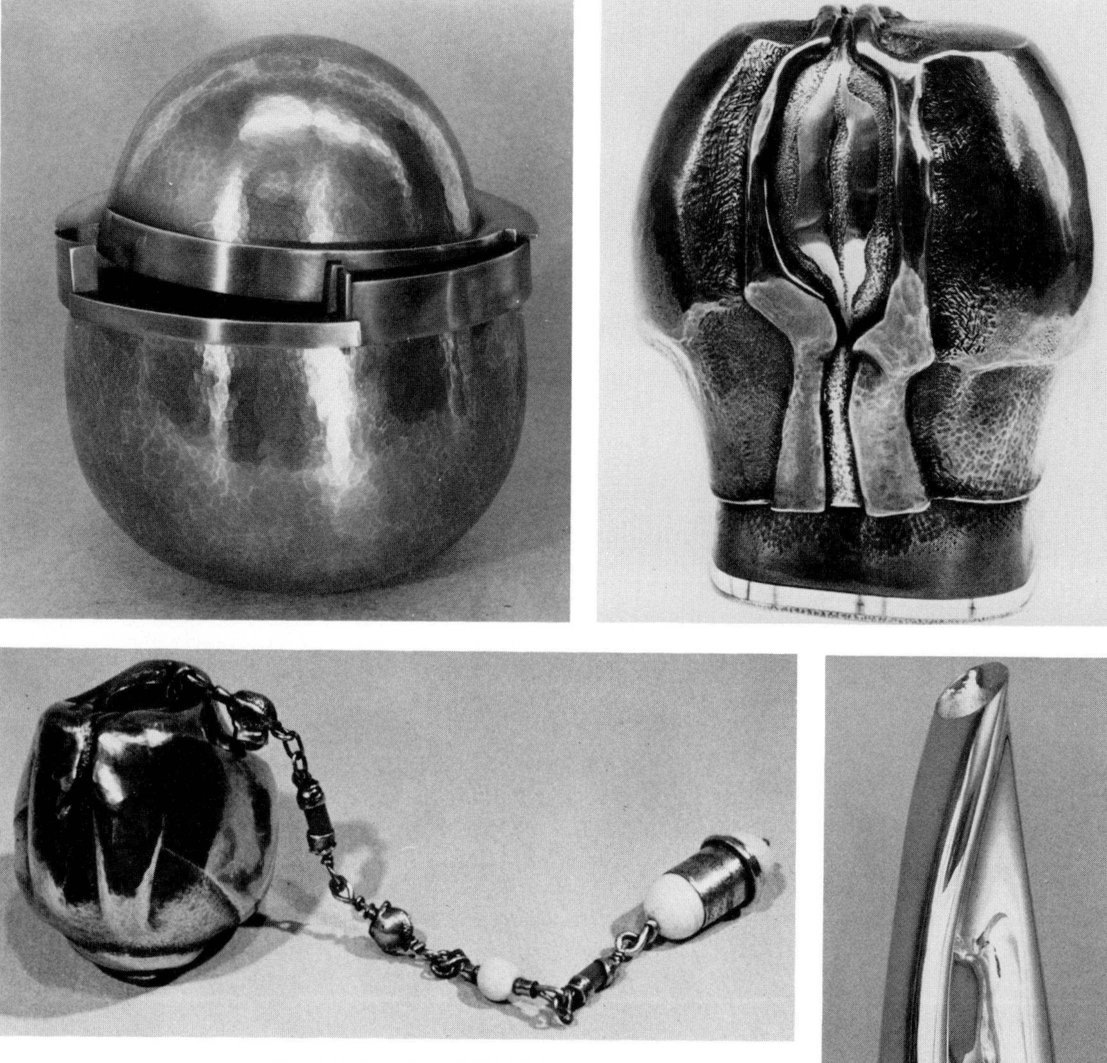

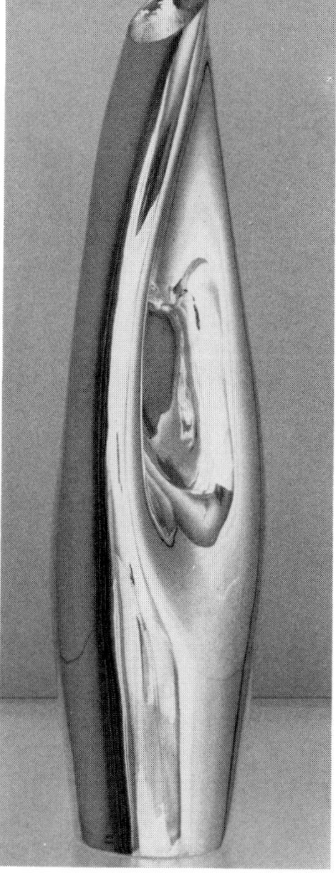

Top, left: Container (16" high) of stretched, crimped, constructed copper, by Ginna Sadler. *Photo courtesy Ginna Sadler*

Top, right: "Tony's Box" (1975) (4" X 3¼"), of bronze, sterling silver, and acrylic, by Barbara Minor. *Courtesy Barbara Minor*

Above: "Worry Box with Beads" (1976) (1½" X 1¾"), in raised and repoussé sterling silver with ivory, coral, and a fabricated chain, by Barbara Minor. *Courtesy Barbara Minor*

Right: View of a raised and chased pitcher (18" high) of sterling silver by Thomas R. Bambas.

Another raised and chased container (10" X 9") by Thomas R. Bambas. *Photos courtesy Thomas R. Bambas*

Fabricating

Fabricating may involve forming as well, but, primarily, it indicates the construction of metal forms by welding and soldering.

The word solder derives from the Latin *solidare,* to make solid, and that is essentially what this attachment technique does. Solder is a metal compound that combines with the metal parts to be attached under heat. Placed on a joint, the solder alloy melts and flows by capillary action into cracks less than 4/1000th of an inch wide. Soldered connections remain secure even when subjected to physical stress—torsional strain and expansion and contraction of metals due to temperature variation and corrosion. At times, the point of attachment can prove to be stronger than the metal itself.

For the artisan, soldering is the vehicle for metal constructions of every size. Copper, brass, silver, gold, platinum, aluminum, can all be soldered. Others, like iron and steel, are best welded.

Procedure of Soldering Steps

The basic steps in soldering are easily mastered:

1. All metal surfaces, including the solder itself, must be perfectly clean. Use a pickle or mild detergent and water, rinse well

2. The solder should be cut into manageable pieces
3. Metal pieces to be joined are fitted together tightly along the joint (joints may be several discrete points or a surface or a line) and, if necessary, tied with iron binding wire to maintain the fit. (But always remove iron binding wire before pickling.)
4. Flux, a chemical that makes the soldered bond possible by removing ever-present oxides from the metal, should be applied along the joint
5. Solder is placed in position on the joint
6. Heat is applied to the metal—*not* the solder. When the piece gets hot enough, solder will flow into the joint. After flow, heat should be removed immediately. (Many torches, including propane, butane, acetylene, and oxyacetylene, may be used.)

Solders should always be chosen for the particular metal to be joined. Since different types of solder have different melting points (and different strengths), experimentation will determine which type is best in any given application and with any given material.

CONSTRUCTED PENDANT CONTAINER by Harlan Butt

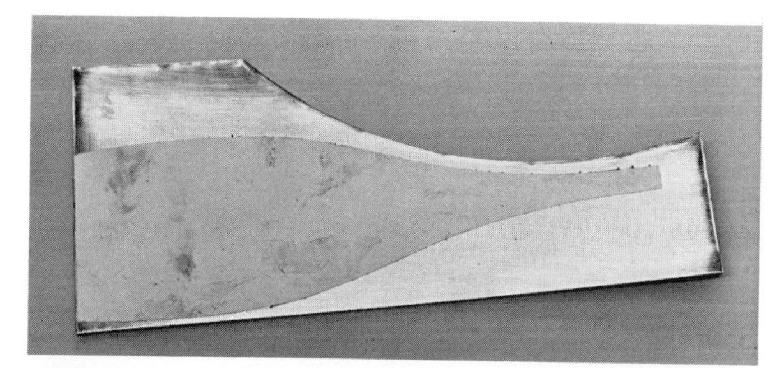

This lizard-pendant container is constructed of formed metal sheet and a cast stopper. Here, the template, which describes the shape of the tail before cutting and forming, is traced onto the metal.

The metal is then cut to size and formed by hammering into a wooden mold.

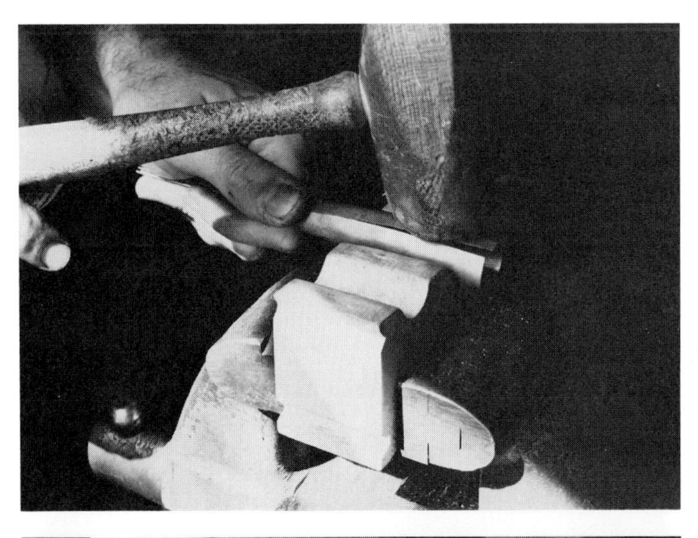

Harlan Butt uses a wooden mallet and forms the metal so that the sides meet evenly at a seam that runs . . .

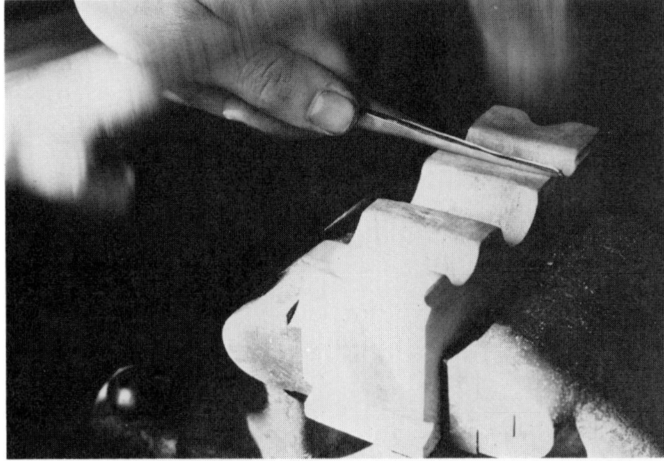

. . . the entire length of the tail. Different spaces in the wood mold accommodate different sections of the form.

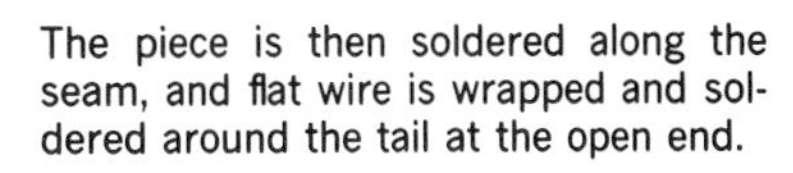

The piece is then soldered along the seam, and flat wire is wrapped and soldered around the tail at the open end.

The lizard head is a casting made from a mold, which was made from a real lizard's head and torso.

The real lizard head and part of its torso are encircled by a cardboard frame made leakproof by clay.

RTV silicone rubber, a flexible and easy to use moldmaking material, is mixed according to its instructions and poured into the frame around the lizard's head. Because RTV silicone flows, it reaches every angle and detail of the lizard's shape, and the mold will allow true to life reproduction of the form.

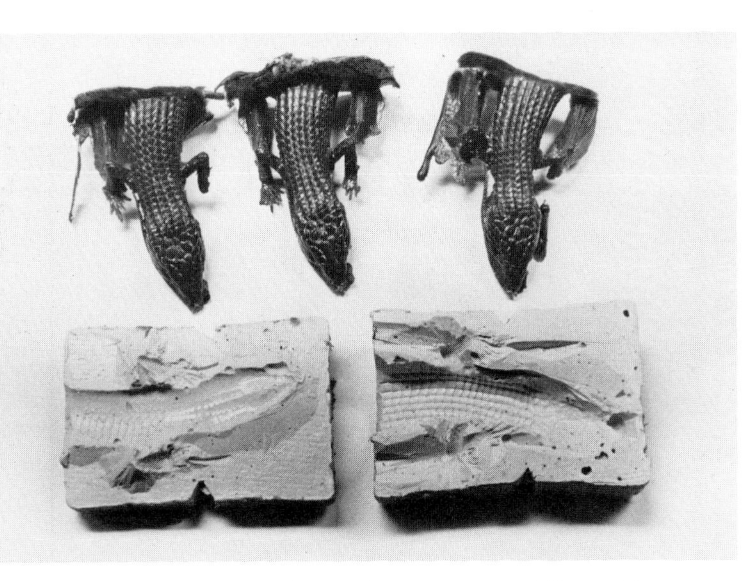

After the RTV silicone cures, the clay and frame are removed. Two notches, one on each side, are made in the mold, in order to achieve completely accurate alignment in castings. The mold is then split, with a sharp knife, into two halves. While some forms may be cast in one-piece molds, the lizard's limbs could not be extricated from such a mold. After the form has been split, the lizard original is removed. The halves are then reassembled, tied into place with notches accurately aligned, and taped securely so that none of the casting resin escapes.

Plastic resin—polyester or epoxy— is then colored if desired, mixed with a catalyst, and poured into the mold. Once the resin has cured, the mold halves are again separated, this time to reveal the finished casting.

The casting may be colored whether the resin was pigmented or not. And the metal tail was also given a rich patina. Constructed metal was attached to the cast lizard head to match and fit it to the tail section.

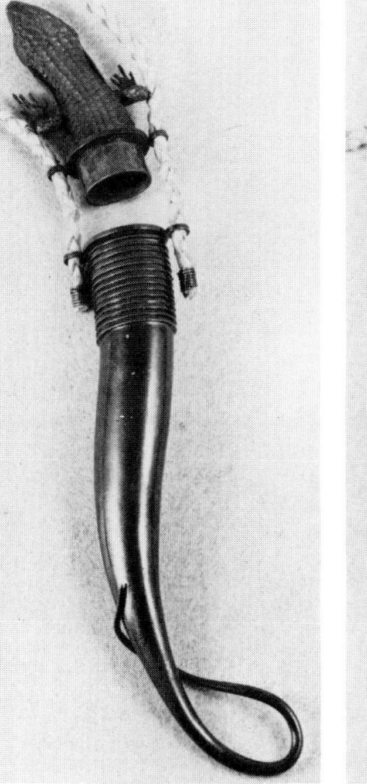

The completed lizard container pendant, by Harlan Butt. *All photos in this series courtesy Harlan Butt*

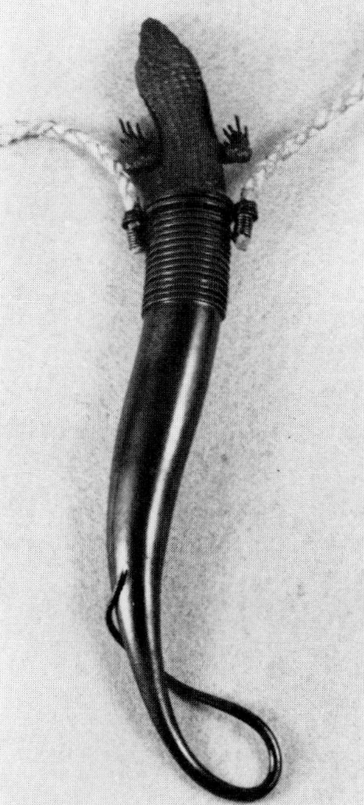

WELDED CONTAINERS by Suzanne Benton

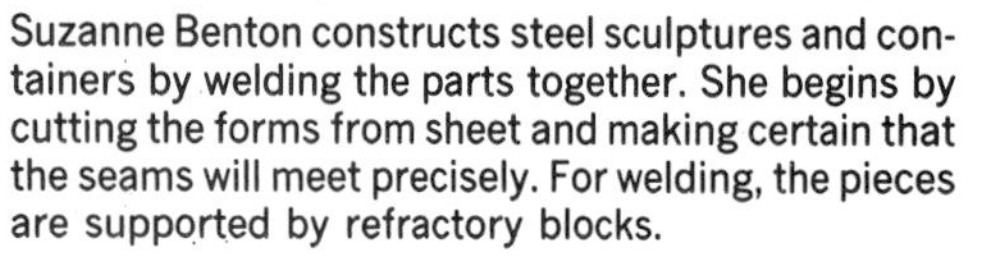

Suzanne Benton constructs steel sculptures and containers by welding the parts together. She begins by cutting the forms from sheet and making certain that the seams will meet precisely. For welding, the pieces are supported by refractory blocks.

During welding, additional support may be applied to achieve a secure joint.

"Prisoner Barred" and "Chain Mask" are contained in a welded steel box, by Suzanne Benton. *Photos courtesy Suzanne Benton,* author of *The Art of Welded Sculpture.* New York: Van Nostrand Reinhold.

"Pot #6" (1972) (24" diam.), of forged and welded steel, by L. Brent Kington. *Courtesy L. Brent Kington*

A formed and constructed kiddush cup (1964) (6" high) by Bernard Bernstein. *Courtesy Bernard Bernstein*

Constructed sterling silver container by Stephen Miller. *Photo by David Luck*

Pewter wine decanter with stainless steel tacks, by Shirley Charron. *Courtesy Shirley Charron; collection of Marvin M. Speiser*

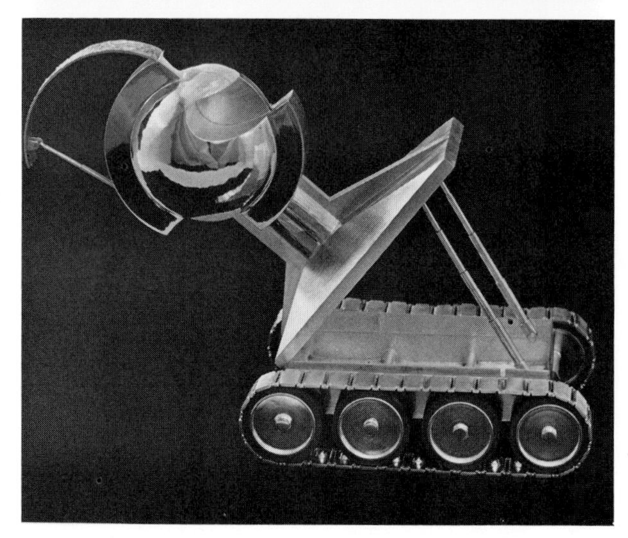

Top: Fabricated two-part fluid bottle (1½" X 5") by Eileen Gilbert Hill. The appendages unscrew at their bases to open. *Photo by Richard Barr*

Above: This silver-plated constructed copper box (6" high), contains another box, by George P. van Duinwyk.

Right: "Outer Worlds and Space Series #4" (1975–1976) (7¾" X 6¾" X 5", 25 ounces). Hal Ross's tracked vehicle is fabricated of sterling silver. It rolls on hand-constructed tracks and wheels . . . and opens in two areas to reveal compartments. Telescoping tube rods lock the lids in both open and closed positions. *Photos courtesy Hal Ross*

Fabricated coffee service of sterling silver and ebony. Coffeepot is 14" tall, by Thomas R. Bambas. *Courtesy Thomas R. Bambas*

Beef bone bottle (4" X 2½") with silver bottom and fabricated silver top. Ebony pins that pull out hold the top in place, by Eileen Gilbert Hill. *Photo by Richard Barr*

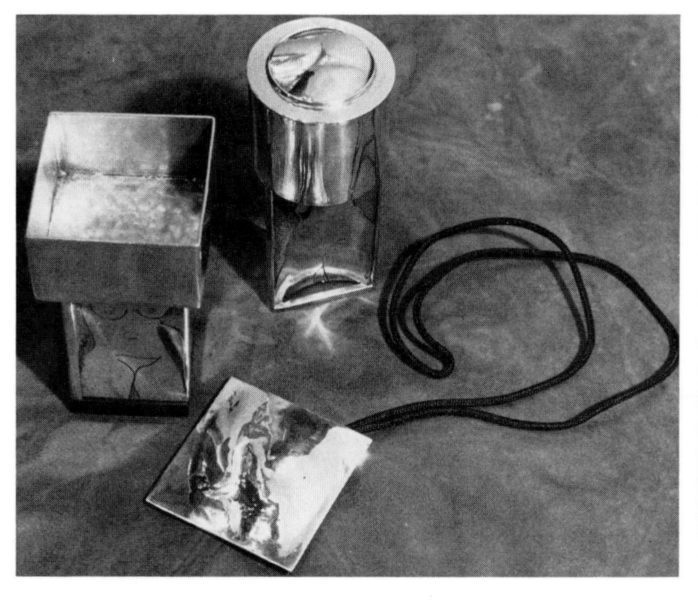

Two constructed silver containers with repoussé and engraving by Stanley and Judith Plotner. Lid of container, *left*, doubles as a pendant. *Photographed at the Kruger Van Eerde Gallery*

A container of shell and fabricated silver by Marguerite Stix. *Photographed at the Kruger Van Eerde Gallery*

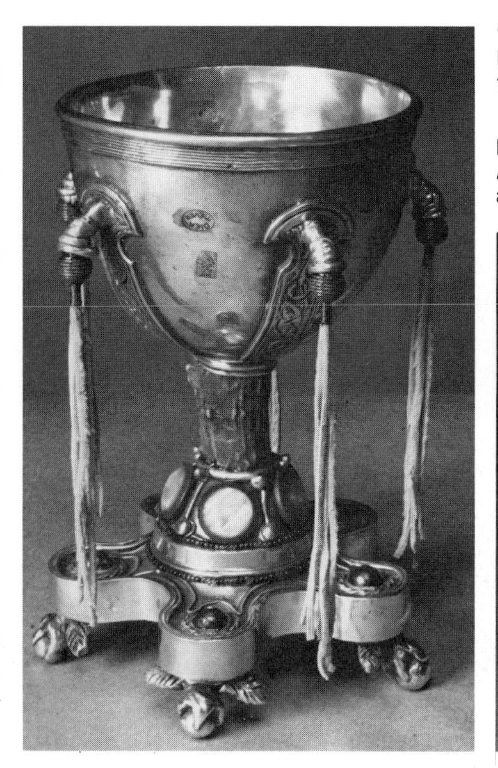

"Renaissance Faire," Pewter, deerhorn stem, streamers of wooden beads, wire, and leather, plastic buttons, and lead pellets in base, brass feet. By Paul A. Diekmeyer.

Paul A. Diekmeyer's "Christian Soldier's Communion Goblet with Field Attachments." Pewter, spike of cow horn, hand guard of human teeth, and attachment of deerhorn. *Photos courtesy Paul A. Diekmeyer*

Harlan Butt's "Hawk Box," constructed of pierced and camphored steel and brass. Hawk's skull lies within.

Right, top: Marcia Lewis's finely crafted "Python Pectoral Purse" (¾" X 8½" X 14") combines brass, mirrored acrylic, python skin, and masonite. *Photo by Dennis J. Dooley*

Right, bottom: "Holster Purse" (32" X 10" X ½"), in brass, resin, and leather, is another carefully designed and skillfully executed work by Marcia Lewis. *Photo by Dennis J. Dooley*

Below: "Feast Box" (6½" high) by Richard Mawdsley. Fabricated sterling silver with cast head, and agate. *Photo by Illinois State University Photo Service*

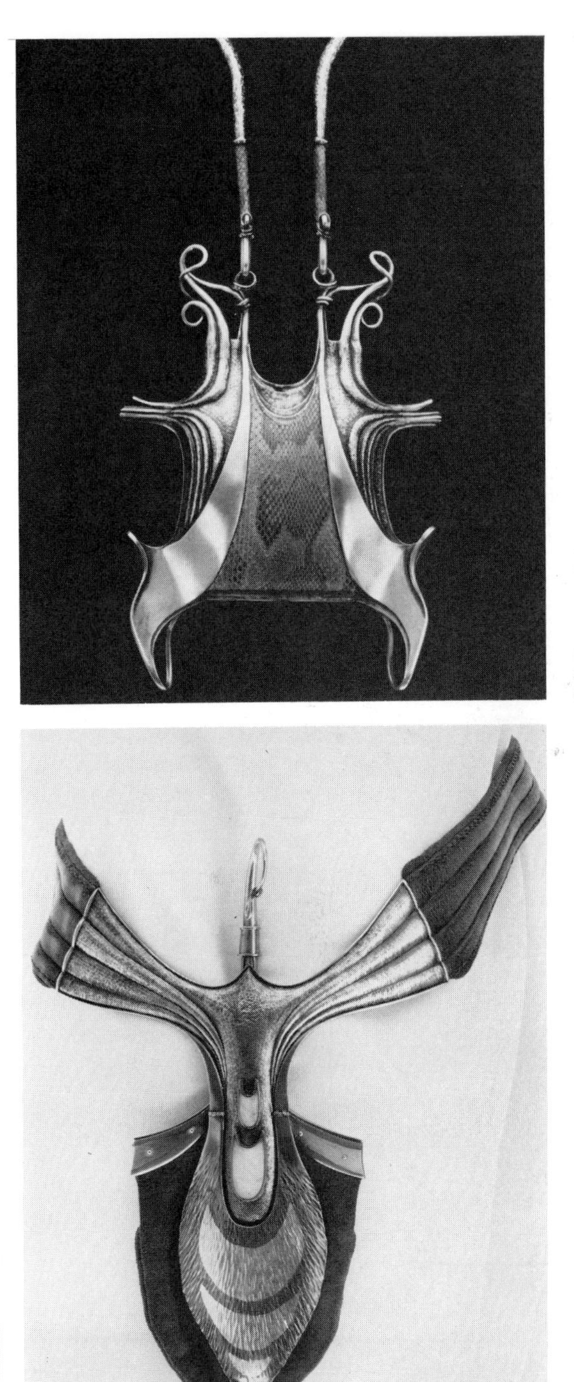

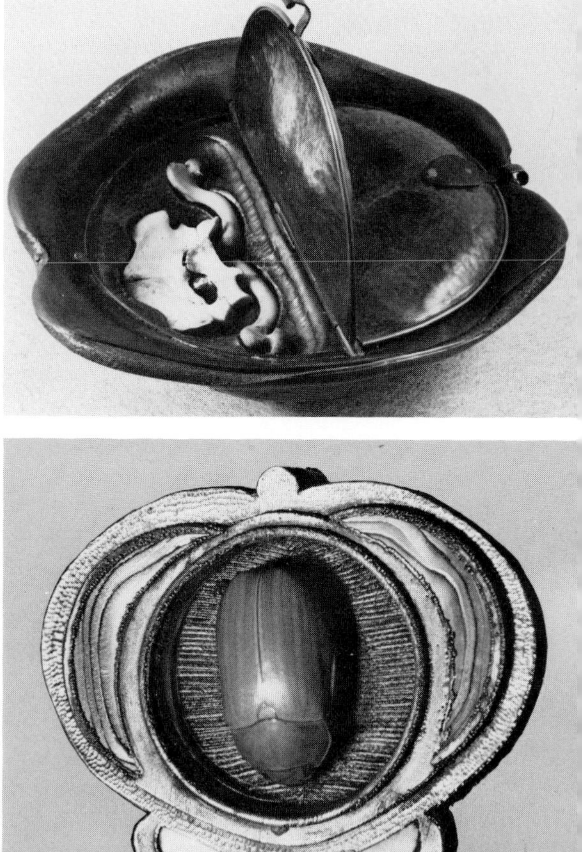

Above: Jamie Bennett constructed this "Steam Cup" (5" X 5" X 3") of silver, carved ivory, carnelian, and 14-karat gold. *Courtesy Jamie Bennett*

Top, right: Reoriented bones of the inspiration lie ensconced within Harlan Butt's construction of raised, chased, repousséd, and pierced copper and steel: "Turtle Box" (1974).

Right: "Entomology Pin #4" is constructed of glass, feather, abalone, mother-of-pearl, copper, and insect.

Below: Jan Brooks Loyd's concern with color and texture is evident in "Entomology Pin #8," a combination of silver, copper, glass tube, padouk, and insect. *Photos courtesy Jan Brooks Loyd*

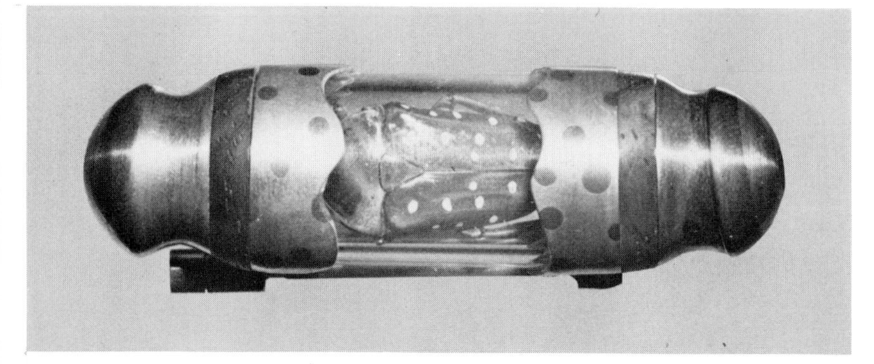

Enameling

Enameling, the combination of powdered glass and metal, is an ancient decorative process. Finely powdered glass—transparent or translucent—applied to the metal, fuses to it when the piece is fired in a kiln. The process may be repeated to build upon previous layers and colors. The result: richly colored forms.

Any of the many enameling techniques can be employed in creating containers. Enamel may be applied directly; partitioned with fine flat wires soldered to the metal surface (cloisonné); fused into spaces in the surface (champlevé); or used as a structural unit, as when units of transparent enamel are supported only by joined wire partitions (plique-à-jour).

Enamels are available in three basic types: transparent, translucent, and opaque. Each type comes in a wide range of colors and degrees of fineness. Some enamels are compounded to be applied to the metal in a dry state, and others react better when wet. The choice depends upon individual preference and the application. Many metals accept enamel readily, but copper, silver, and gold are used most often and all work well. No matter which metal is used, however, the surface must be thoroughly cleaned first. The surface must be degreased and deoxidized in a "pickle" of dilute acid or a nonacid metal cleaner.

After cleaning, an adhesive solution, which will hold the glass particles to the metal during firing, is sprayed or brushed onto the surface. Lavender oil or gum solution is formulated for just this purpose.

Usually the metal piece is then placed on white paper to capture any enamel that spills and thereby avoid waste. Enamel may then be applied.

The fine glass may be sifted or sprayed onto the metal. It may be applied with a brush, spoon, or needle. Or it may, depending upon the formulation, be applied wet or dry. The mode depends upon the desired result.

Once the enamel has been applied, the piece is ready to be fired. During firing, the form may be placed upon a metal enameling rack, ceramic trivet and nichrome wire, or nichrome wire inserted into refractory blocks, and then placed within the enameling kiln at 1500°F. Strict attention must be paid to the form during firing, and the piece is removed once the enamel looks smooth and glossy.

Once cool, oxidation is cleaned away and firing may be repeated until the desired effect has been achieved.

Finally, enamels may be finished by removing rough or blackened edges with a file or stone and refined with emery cloth and steel wool. Uneven surfaces may be further smoothed and leveled by grinding them with carborundum stone under running water. The surface may be left matte, as some enamelists prefer, or brought to a glasslike gloss with rouge.

CLOISONNÉ CONTAINERS by Marian Slepian

Marian Slepian creates enameled containers and container tops. The first step in creating an enameled top for a wood container is to measure 20-gauge copper and mark the area to be used. The copper is cut using wire shears, and the fit is tested. Because the copper expands under heat of firing, 1/16" leeway is allowed.

The edges and corners are filed off; otherwise the stress causes the enamel to crack. The edges are hammered to flatten and straighten them. Both sides of the copper are then pounded with a rubber mallet to eliminate unsightly mill roll.

The surfaces are then cleaned with scouring powder and steel wool, and the metal is rinsed thoroughly. The back of the copper is then prepared for counterenameling. By enameling the underside as the top is enameled with the desired design, the stress is equalized, preventing cracking of the top surface. Any time that enamel is applied to one side, it also must be applied to the other side. Before applying enamel, the bottom surface is sprayed with a solution of 2 parts water to 1 part gum tragacanth. This binds the enamel to the surface and allows both sides to be enameled at the same time. The gum tragacanth burns away during firing. However, if too much gum is used, it leaves a shadow on the enamel.

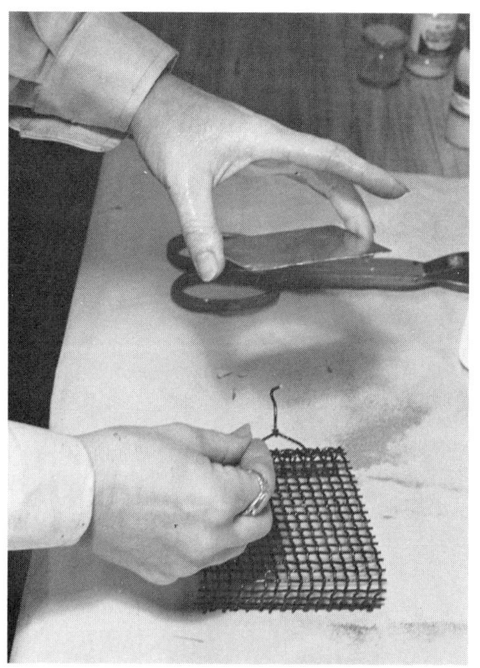

The cleaned and sprayed copper is then placed upon a trivet of nichrome wire, which has been placed on a stilt.

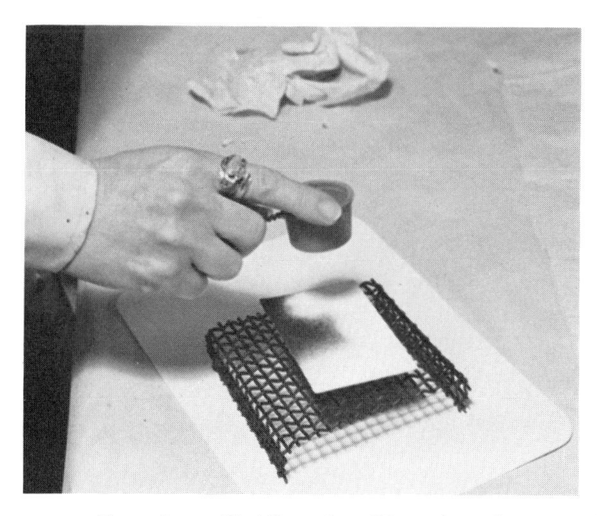

Enamel was sifted through an 80-mesh strainer onto the bottom (counterenameled) side, and a coat of clear flux is sifted onto the top surface as a base for future layers of enamel.

The fluxed and counterenameled piece is then fired.

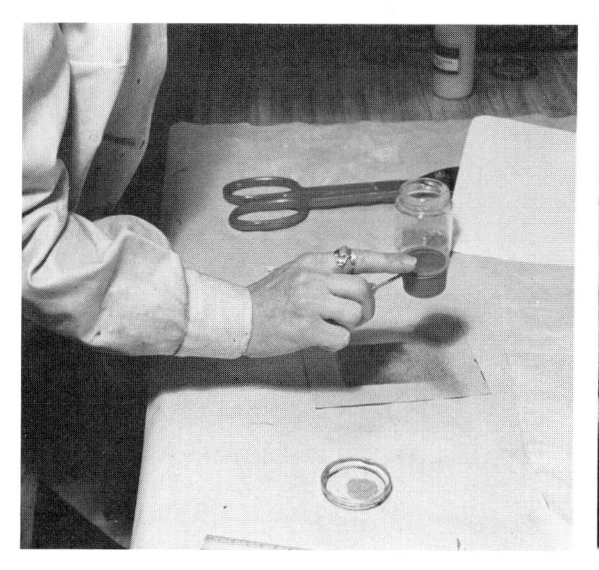

A base layer of colored enamel is then applied and fired.

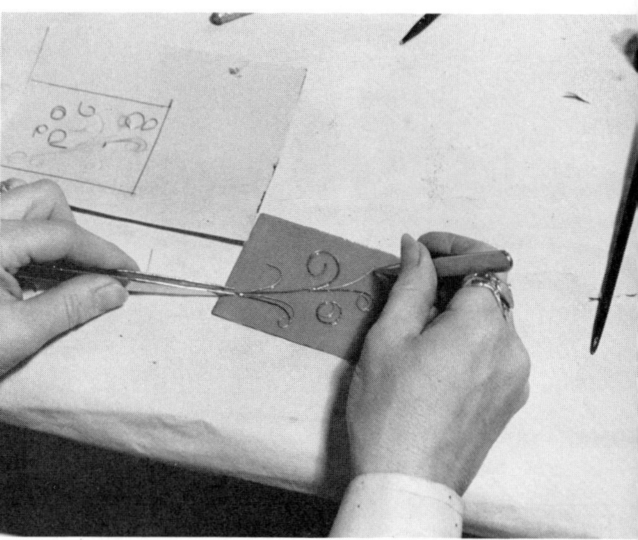

The cloisons here are of thin round copper wire and flat silver wire. The wires are annealed first to make them soft and easily shaped. Annealing is achieved by placing the wires in the kiln until they turn red. They are removed and cleaned in pickling or Sparex®. After annealing, the wires are shaped with tweezers and pliers to match Marian Slepian's pattern. The wires are pressed until they sit flat, dipped in pure agar gum, and placed on the twice-enameled copper. The agar is allowed to dry.

Clear flux is then sifted over the wires and fired for just a minute, until the flux sets and no longer looks like powder.

The briefly fired copper with cloisons is then weighted under the iron until it cools, to prevent curling.

The enamels are then mixed into a suspension of lavender oil or water. Enough of the liquid is added to create a mixture the consistency of sour cream.

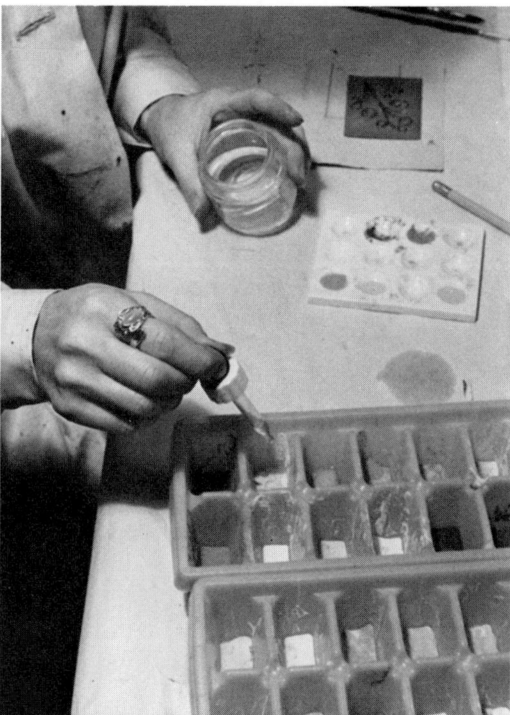

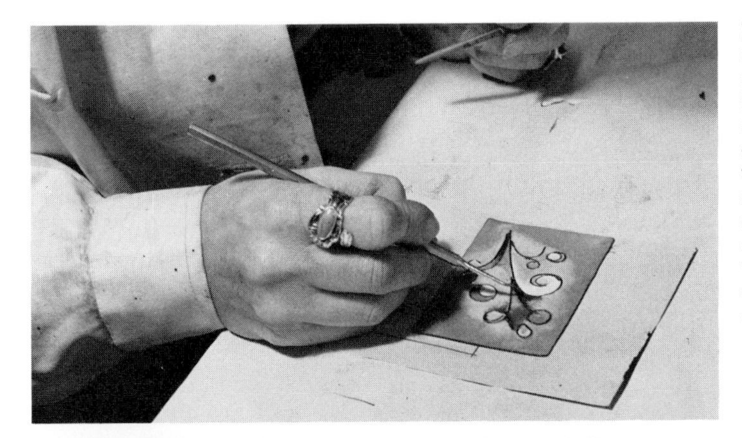

Enamels may be applied by any convenient means; Marian Slepian uses a paintbrush. Colors that will not burn easily should be applied first, and opaques should be applied before transparent colors. In each firing, however, it is important to treat each part of the surface exactly alike. If the design does not call for color at a particular level in the enamel, clear flux is applied. Always cover the entire surface to the same thickness. To do otherwise would cause cracking and uneven buildup.

After each application of enamel the piece is fired until the surface is glossy. The process is repeated until the enamel reaches the top of the wires.

Once the enameling has been completed, the surface may be leveled by grinding with carborundum stone. Grinding should be conducted under running water, and should progress from coarse to fine stones. After grinding, the form may be flash-fired—fired for a few seconds—to bring back the gloss, or waxed and buffed for a matte finish.

Above: The enameled plaque is glued to the box inset with Elmer's Heavy Grip, a contact cement. A weight is used to assure a firm bond.

Right, top: The finished cloisonné-topped container by Marian Slepian.

Center: Another design in cloisonné by Marian Slepian.

Below: A brass container with cloisonné top by Mary Kretsinger.

Black lacquered wooden boxes (8½" X 8½" X 3") with covers inset with transparent enamel on copper, by Saara Hopea-Untracht. *Photo by Oppi Untracht*

Enameled silver jewelry box by Joyce Mills. *Courtesy Buckinghamshire College*

June Schwarcz and Imogene Gieling collaborated on this silver Kiddush cup for the Temple Emanuel of Denver, Colorado. The bowl was hammered, etched, and enameled by June Schwarcz. The silver- and gold-plated base was cast and forged by Imogene Gieling. *Courtesy June Schwarcz*

A covered copper bowl, with brass lip, in orange enamel with flux-covered oxides, by Oppi Untracht. *Photo by Oppi Untracht*

Electroforming and Electroplating

Electroforming and electroplating are techniques that are just now being employed by artists. For this part of the book, we owe a great debt to Timothy M. Glotzbach, William O. Huggins, and Joseph S. Clift, students of L. Brent Kington at Southern Illinois University. They compiled and wrote most of this section. We feel that craftsmen and artists will find this exploration of the basic processes, without much of the highly technical information that dominates the literature, extremely easy to apply. The copper-acid bath/D.C. rectifier system illustrated here is one popular setup; other systems, solutions, and technical information are contained in the books listed in the bibliography.

The electroforming process is illustrated by a series of photographs showing the creation of a container by Tim Glotzbach.

David Luck, a student of Julius Schmidt and Chunghi Choo, at the University of Iowa, also illustrates the electroforming process.

One word of caution: Always observe the utmost care when working with the chemicals involved in electroforming. Some are highly caustic; always wear gloves and flush with water after any contact with an acid. Consult a physician for further care if any indication warrants.

Electroforming and Electroplating: Techniques and Application

Electroforming and electroplating have a history of industrial development and application dating back to the mid-nineteenth century. Only in recent years, however, have artists and craftsmen working in metal made extensive use of these processes as structural and decorative media.

Electroforming and electroplating are methods of covering objects with a layer of metal by means of electrodeposition. If the object is permanent, the process is electroplating and is, essentially, the application of a thin layer of metal to an otherwise finished form. This may be done to achieve a more durable or tarnish-resistant surface or one of a different color or texture. Electroplating of ceramic, stone, metal, plastic, even fabric objects, is possible.

If the original form is removable, the process is electroforming and the object is the matrix. Electroforming allows the creation of articles that are hollow, light, and difficult or impossible to create by any other means. Intricate shapes made of wax, for example, can readily be translated into metal by electroforming, combining the versatility of one medium and the surface qualities of metal. In this example, the original wax matrix does not remain part of the completed form.

Combinations of the two processes are also possible. In other words, the deposits may be applied to an object that is part metal and remains in the completed article and part matrix that may or may not be removed (as is demonstrated by the accompanying photographs). Electroforming may also be used as a joining technique or to create and enrich surface texture.

These processes are unique. The metal is deposited without high temperature, making it possible to include various types of materials in your work (bone, plastic) that do not lend themselves either physically or aesthetically to typical joining techniques such as soldering, bezels, or riveting. The resulting forms are extremely light for their size, despite the fact that often they appear to be extremely heavy.

In both electroforming and electroplating, the essential process is electrodeposition of metal. Simply put, the plating metal and the object to be formed or plated are placed in an acid bath; the object is connected to a negatively charged source of electricity, while the metal is given a positive electrical charge. Since the negative and positive charges attract, metal ions flow toward the object and are deposited on its surface. More precisely, current travels from the positively charged anode (source of metal) to the negatively charged cathode (the object to be plated) through the electrolyte (plating bath or solution). The anode frequently serves as a continuing source of metal ions for deposition, and these ions are carried through the electrolyte to the cathode but, in some cases, the anode is insoluble. The bath itself is then the source of ions and must necessarily be replenished from time to time. The longer the matrix remains in the bath, the thicker the electroform or plating will be.

The Matrix

The matrix on which you electroform can be metallic, nonmetallic, or mixed metal and nonmetal. An all-metal matrix is probably the easiest to electroform, assuming that the matrix metal is compatible with the electroforming bath. For example, the sulfuric acid in a typical copper bath will etch mild steel, making it difficult to achieve a strong deposit of copper on this metal. All that is necessary to prepare a metal matrix for electroforming is careful attachment—by soldering, gluing, or wrapping—of the cathode lead wires and thorough cleaning of the matrix to remove grease and oxidation.

Nonmetallic matrixes are slightly more complicated to electroform because they must first be made electrically conductive. This is usually achieved by degreasing and then painting with an electroconductive paint. (Two good silver-base paints are Electrodag 416, Acheson Colloids Co., Port Huron, Michigan 48060 and Dupont Conductive Silver, Burt Bricker, Inc., P.O. Box 171, Wilmington, Delaware 19899.) Attaching the cathode wires to a nonmetal matrix is often difficult, but it can be simplified by designing a mechanism into the article that will facilitate attachment. Mixed metal and nonmetal matrices are often a solution to this problem of cathode wire attachment. Metal, if placed strategically, can also provide reinforcement for the electroformed object. Again the piece must be thoroughly cleaned of all grease and oxidation, nonmetal areas must be made electrically conductive, and cathode wires must be attached (to a metal armature if possible.).

Once electroforming has been completed, it is often desirable to remove the matrix, particularly if it is a substance such as wax. This can be easily achieved if a small opening is allowed to remain in the finished object. (This opening may be hidden from view.) The wax may be boiled out in soapy water, dissolved with chemicals, or burned out. Your choice of matrix will determine the possibility and technique of removal. However, if using heat, be careful not to totally anneal your piece.

The copper is deposited in an unannealed state, which accounts for the strength of the very thin form. Any hard soldering or equivalent heat will soften the electroform enough to undermine its structure. The only way to achieve work hardening is to reelectroform the object.

When choosing a matrix, it is important to select a material that will not be destroyed by the electroforming bath. (Bone is attacked by acid.) Organic matrices should be given a protective coating such as spray plastic or lacquer. Problems will also be minimized by planning all procedures in advance.

Keep in mind such factors as desired strength in finished object, whether the matrix will be easy to remove, whether it is possible to leave the matrix in place, and whether the matrix will remain intact throughout the process.

In order for an electroformed/plated deposit to adhere securely, the surface to be formed or plated must be completely clean. A simple way to clean the matrix or object is to scrub it with a liquid detergent and a fine abrasive such as pumice. It is also possible to degrease the matrix in an ultrasonic cleaner filled with a mild ammonia solution. A third cleaning technique (for metal matrices) is electrocleaning. This is accomplished by immersing the matrix in a commercially prepared electrocleaning solution contained in a stainless steel tank. The positive lead from the rectifier is attached to the matrix, the negative lead meets the stainless tank. Current and temperature of the cleaning solution are adjusted according to the manufacturer's instructions. Metal matrices can also be cleaned of grease and oxidation by pickling. After all cleaning processes, the matrix should be rinsed thoroughly in tap water and then in distilled water before immersion in the electroforming bath. Of course, the matrix should never be touched after cleaning, since oil from your skin will contaminate it and retard electrodeposition.

Control of the Process

After the initial electroforming, it may become apparent that parts of the form are growing too quickly, while other parts are not growing fast enough. Repositioning the piece in the bath often remedies this uneven growth. Face slow-growing areas more directly toward the metal anodes and give the fast-growing areas less exposure.

Two other ways to effect and control growth in special areas are stopping out and shielding. Stopping out is accomplished by painting a resist (a nonconductive substance) directly on the surface of the matrix. This prevents metal deposition in the painted areas. Many substances can be used as resists; wax, rubber latex, lacquer, and spray plastics are effective and easy to remove. A commercial product called "Micro Peel" (Michigan Chrome and Chemical Co., 8615 Grinnell Avenue., Detroit, Michigan 48312) is very effective and versatile and can be removed with acetone. Of course, if the matrix is nonmetallic, you can control the areas of deposition by selectively painting with the electroconductive paint.

Shielding to prevent growth is accomplished by placing a sheet of wax or similar nonconductive material above the matrix and between the matrix and the anodes. This prevents rapid deposition of metal on the matrix in the area shielded from the anode surface.

When choosing a resist technique, consider that when areas of the matrix are stopped out, a hard raised edge is usually visible on the finished object where the resist ended and electroforming began. If this hard raised edge is not aesthetically acceptable, use the shielding technique. The shield does not completely stop growth but severely retards it in the area protected from anode exposure; the result is a softer, more gradual line.

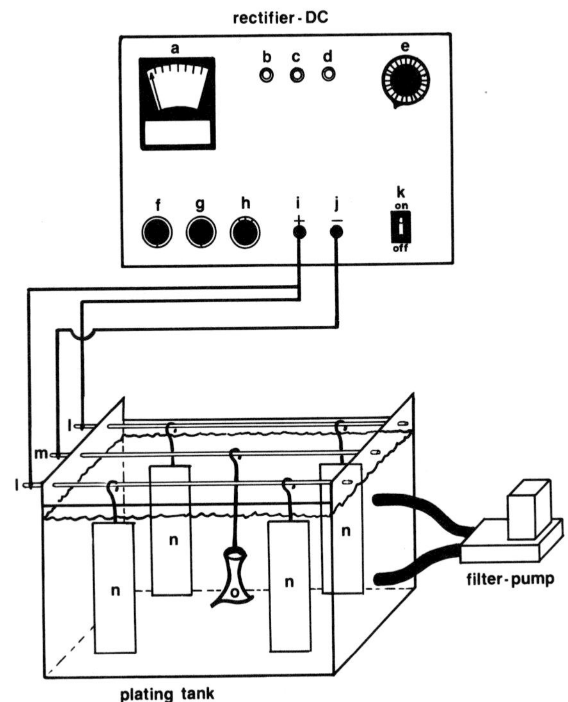

plating tank
(front wall removed to show anode bars and cathode)

Key: a) ammeter, b) forward indicator light, c) reverse indicator light, d) on-off indicator light, e) rheostat for current adjustment, f) forward plating time control, g) reverse plating time control, h) switch (forward, reverse, cycle), i) positive electrode connection, j) negative electrode connection, k) on-off switch (rectifier), l) anode bars (positive), m) cathode bars (negative), n) anodes (plates: source of metal ions), o) cathode (object).

The system shown here is optimal, but other systems are possible. Batteries of varying sizes may be substituted for this power control source; however, this will require experimentation and careful supervision. Depending upon the size of the tank and the desired current density, a car battery, dry cell, model train pack, or battery charger might be effective.

If only electroplating and not electroforming, it is possible to dispense with the filter—although it still helps to have it. And it is also possible to use fewer metal anode bars of copper and still achieve good results, although it may be necessary to turn the matrix since the metal ions will not be as evenly distributed. Agitation, while always beneficial, is not as critical in electroplating as it is in electroforming.

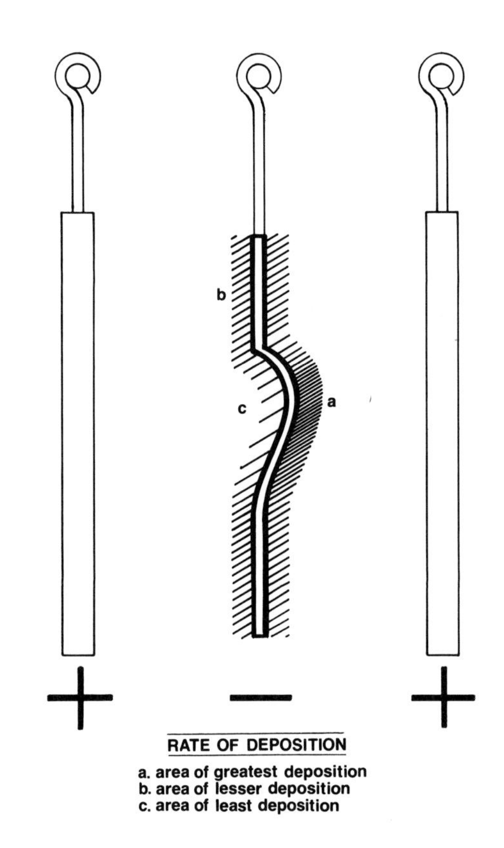

RATE OF DEPOSITION

a. area of greatest deposition
b. area of lesser deposition
c. area of least deposition

Working Vocabulary

Before examining the actual steps in electroforming, consider this working vocabulary of electrical terms that apply to electrodeposition:

Current: The electrical charge flowing past a specified circuit point per unit time. Current is measured in amperes (amps).

Amperage: The measure of current or the electrical charge.

DC (direct current): Current that runs from + to – only.

Current Density: The ratio of the magnitude of current flowing in a conductor (solution) to the size of the cathode (object). This ratio is always proportional and is measured as amperes per square foot of cathode surface ($amps/ft^2$). It is the critical determinant of the speed of electroplating/forming.

Anode: Any positively charged electrode, in this case the metal plates used in the electroforming tank.

Cathode: Any negatively charged electrode, in this case the work or piece to be electroformed/plated.

Deposition and Rate of Deposition: The actual metal coating being deposited on the piece and the rate of coating per unit time.

Periodic Reverse: This system is built into a rectifier and can be activated when necessary. This method, often utilized in plating and electroforming, purports to achieve a uniform deposition of metal by changing the direction of current flow periodically (i.e., cathode becomes anode, and anode becomes cathode). The duration of this change is controlled by two timing devices built into the rectifier. When setting these timers, keep in mind that the amount of reverse stripping time must be less than half the amount of forward plating time in order to achieve deposition (see accompanying diagram). It should be noted that many feel this system is unnecessary, and there is controversy over whether this system actually achieves its purpose while using the copper-acid bath.

Solving Electroforming Problems: Glotzbach, Huggins, and Clift

PROBLEM	SOLUTION
No current or current drops off.	Check the connections of both anode and cathode. Is there sufficient agitation? Check the wires running from the rectifier to the tank. Check the temperature of the bath. Check for proper operation of rectifier. Is the bath properly prepared?

PROBLEM	SOLUTION
Certain areas are not plating.	Was the matrix clean before plating? Are all nonmetallic areas thoroughly coated with a conductive solution? Was the conductive solution mixed thoroughly? Is it too old?
Parts of the plating peels from the matrix or brushes and pops off.	Was the matrix clean before plating? Is the metal being deposited too quickly to adhere properly? (Too much amperage can actually burn the metal deposited.) Check for correct current setting.
Certain areas build up faster than others.	These areas are too close to the anode. Do these areas need to be masked out or shielded? Is there a need to reposition the anode bars or reposition the cathode?

Finally, there are several factors that are particularly important to the electroforming process. They should be kept in mind at all times because of their importance in the success of the process:

Current: The current flowing through the electrolytic solution is important for surface buildup. A large cathode will require a higher current to deposit metal at a convenient rate. Thus, change in the amount of current (amperage) flowing through the bath will change the rate and character of deposition. Different currents may be employed to achieve variety in surface treatment. Initially, it is best to follow the current settings recommended for the bath being used. Experimentation will allow adjustment of current for particular applications.

Heat: Most solutions, commercial and homemade, operate best within a particular temperature range. Heat aids in keeping the salts of the solution dissolved and the ions free to move in the electrolyte. Heat can be administered through the use of an aquarium heater in the tank or the solution may be heated on a hot plate prior to or during use. Again, It is best to maintain the temperature range recommended for the particular bath.

Agitation: Agitation, or stirring of the solution, keeps a fresh supply of metal ions in movement around the cathode and in the solution. A lack of agitation causes the immediate area around the cathode to be depleted of positive ions, thus decreasing the efficiency of metal deposition. Agitation also sweeps away gas bubbles from the cathode surface, thus preventing pitting. (Pumps and filters are manufactured by Serfilco, 1415 Waukegan Road, Northbrook, Illinois 60062. Plating supplies are available from the Udylite Corp., Detroit, Michigan 48234.)

Operating Conditions for Electroforming

COPPER PLATING

Copper sulfate bath: copper sulfate—32.0 oz./gal. distilled water
sulfuric acid—8.0–10.0 oz./gal. distilled water
Anode metal: pure annealed copper (stripped and cleaned)
Temperature: 60–120°F.
Voltage: 1–4
Current density: 20–50 amp/ft^2 (normal) (dependent on tank size)

SILVER PLATING

Silver bath: silver cyanide—5 oz./gal. distilled water
potassium cyanide—8 oz./gal. distilled water
potassium carbonate—2 oz./gal. distilled water
Anode metal: "Three-nines" silver (99.999% pure)
Temperature: 60–120°F.
Voltage: 1–4
Current density: 20–50 amp/ft^2 (normal) (dependent on tank size)

Procedures for Electroforming

These are the basic steps in electroforming:

1. Prepare the matrix
2. Attach the cathode wire in an appropriate place
3. Clean the matrix. If the matrix is metal and areas are to be stopped out, it should be done now. Do *not* touch surface
4. Make the matrix electrically conductive if it is not metal
5. Attach to cathode bar above tank and place the matrix in the bath. An alligator clip works nicely to hold cathode wire to bar
6. Turn on rectifier and set the current low until the matrix is totally pink. If the current is set too high, it may burn the conductive paint or interfere with proper adhesion of the plating
7. Turn up amperage. Leave amperage up as long as you intend to plate. If you are uncertain about the length of time to electroplate/electroform, check the piece periodically. Also check for uneven growth

8. Remove the form from the bath, rinse it in water, and neutralize it in baking soda and warm water
9. Remove stop-out, if any was used
10. Remove matrix, if necessary
11. Remove the cathode wires. It is possible to reelectroform to cover the area of the cathode connection
12. Finish piece as desired (patina)

Electroforming Sequence

The accompanying photographs show the steps in constructing an electroformed copper container. It is composed of two matching permanent armatures with small sections of 1/8" copper tubing soldered to them to serve as a hinge in the finished piece. The armatures provide a framework on which a wax matrix may be modeled. After a thin layer of copper is electroformed on the matrices, the wax is removed by immersion in boiling water. Further forming is done to add strength to the piece. This subsequent forming is done on the inside in order to cover the silver gray color of the Electrodag (which remains after the wax matrix is removed) and to preserve the detail of the outside texture that resulted from the carving of the wax. The hinge and other areas that fit closely together must be stopped out with wax or lacquer to prevent the forming of metal that would interfere with the functioning of the container. While the interior is strengthened, the exterior receives decorative textural plating. This is accomplished by stopping out most of the surface. Only limited areas, which emphasize the edges of the forms, are left exposed, and they are plated at a higher amperage in order to create a grainy surface. As a part of this decorative plating, the cement that holds a small piece of enameled copper on the top is also covered with a linear pattern of grainy textural plating. After all plating and forming are done, the piece is given a deep gray black patina by dipping it in a hot concentrated solution of liver of sulfur in water. The darkened surface is highlighted by rubbing with moist, fine pumice powder. This removes some of the dark color. After rubbing carefully with steel wool to brighten the surface, paste wax is applied to the entire piece and it is hand-buffed with a shoe brush to give it a soft sheen.

ELECTROFORMED CONTAINER by Joseph C. Clift

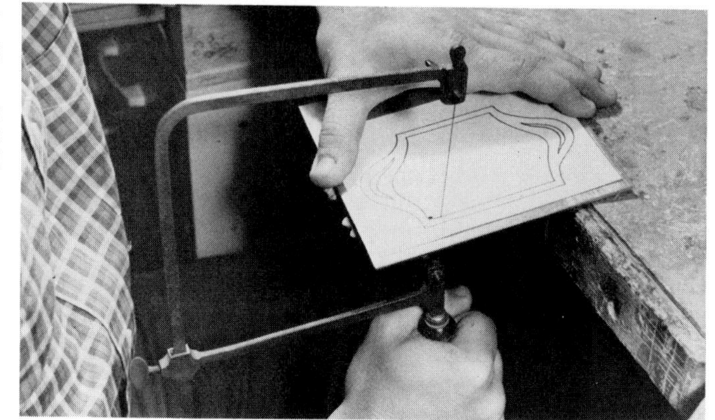

One of the two armatures is cut from 16-gauge copper sheet with a jeweler's piercing saw. A paper pattern cemented to the metal aids accurate sawing.

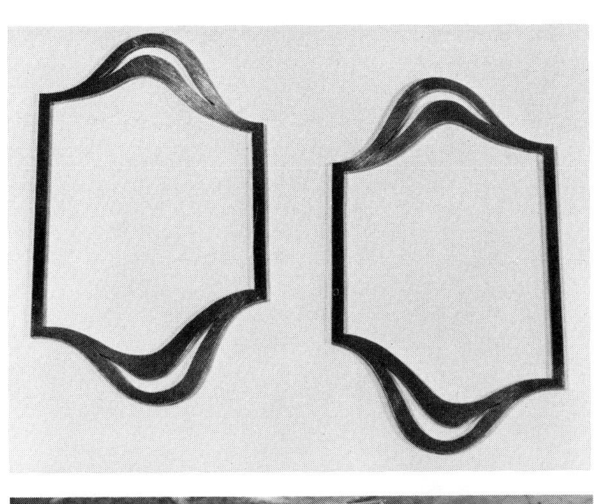

The two armatures are cut out and all rough edges are filed and sanded.

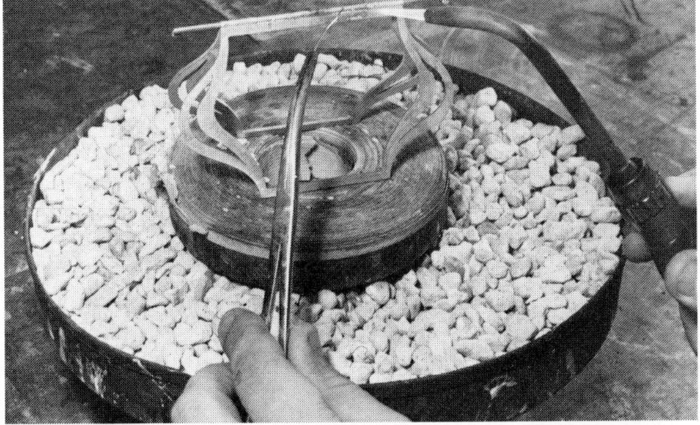

The sections of tubing that will comprise the hinge are silver-soldered into place. A piece of wire running through the hinge keeps the individual pieces aligned. This step must be done with great care to avoid flowing solder into the wrong places and "freezing" the hinge into a solid piece.

Wax is attached to the armature and modeled to the desired form. This piece will be the top half of the container.

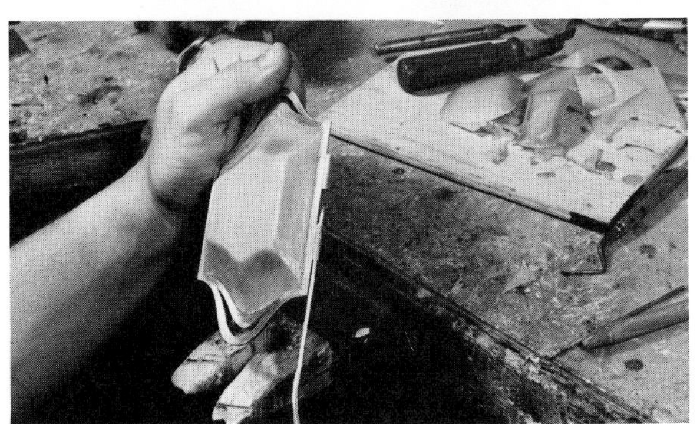

The electrode wire (10-gauge copper) is securely attached to the top armature. The hinge tubing provides an excellent point of attachment in this case. Wax is used to protect the hinge area from unwanted metal formation.

Wax work and electrode wire are attached to the bottom armature as well.

Electrodag (electroconductive paint) is applied to the top armature. The coating must cover the wax matrix and extend slightly onto the copper armature. This assures a firm bond between the newly formed metal and the existing metal of the armature.

The bottom piece, having been painted with Electrodag, is placed in the tank for forming. It must be securely attached to the cathode bar in the center.

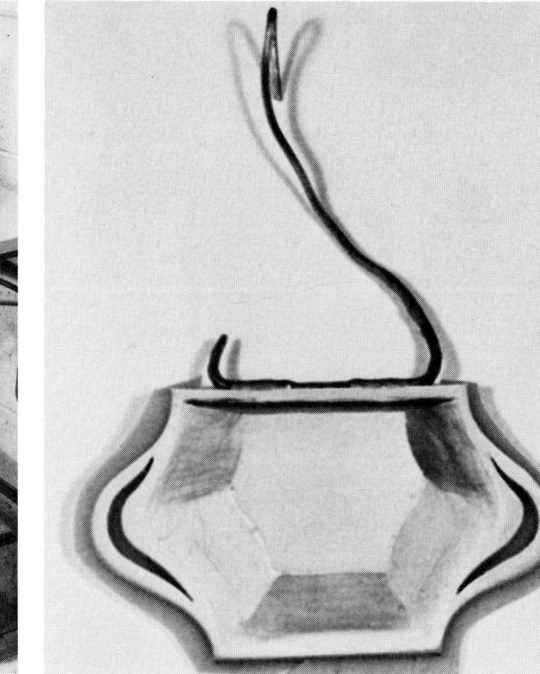

The first thin layer has been formed and the wax may now be boiled out. The plated metal should be a uniform light copper pink color.

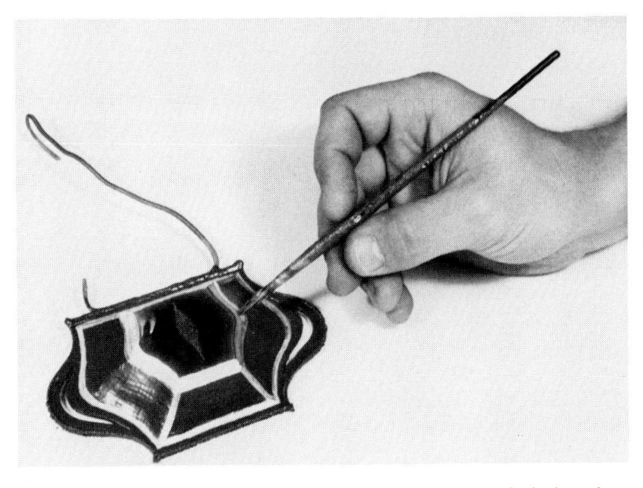

The wax has been boiled out, and the enameled piece has been cemented in place. Certain areas have been stopped out with lacquer (dark areas). Electrodag is painted over the cement and very slightly up onto the surface of the enamel. This will cause the enamel to be held in place with an electroformed bezel.

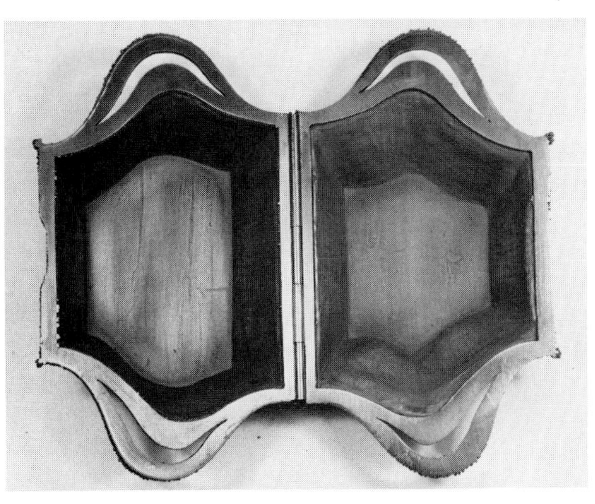

This is the interior of the completed container. The hinge pin has been inserted and the gray black patina applied.

Top view of the finished piece.

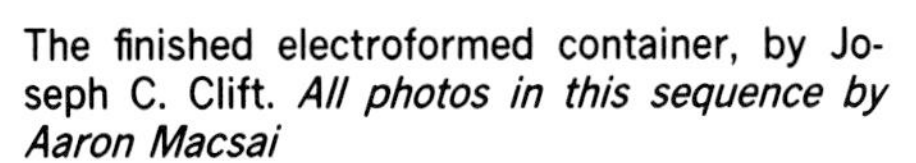

The finished electroformed container, by Joseph C. Clift. *All photos in this sequence by Aaron Macsai*

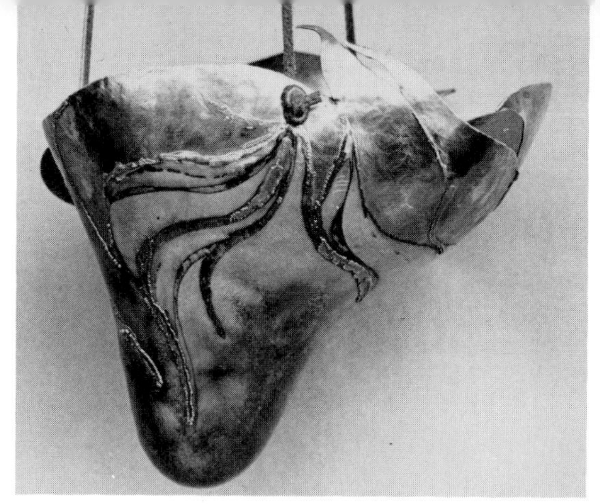

"Spring Planter," by E. M. Klim, of electroformed, enameled copper. *Courtesy E. M. Klim*

Foil bowl, electroplated and enameled, by June Schwarcz. *Courtesy June Schwarcz*

ELECTROFORMING OVER WAX MATRIX

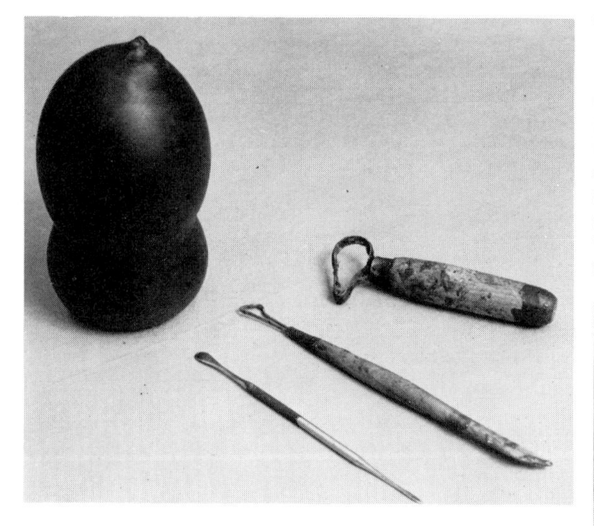

This electroforming sequence was photographed by David Luck at the University of Iowa, under the supervision of Professor Chunghi Choo of the Art and Art History Department. Here a typical hollowware form has been readied for plating. In this case, microcrystalline wax is being used as the matrix, although other materials may also be used. The consistency of the wax may be controlled by varying the amount of paraffin that is added. Wax working tools, some of which are shown, range from dental instruments to ceramic tools, to scouring cloths, and wax solvents used to achieve a smooth surface.

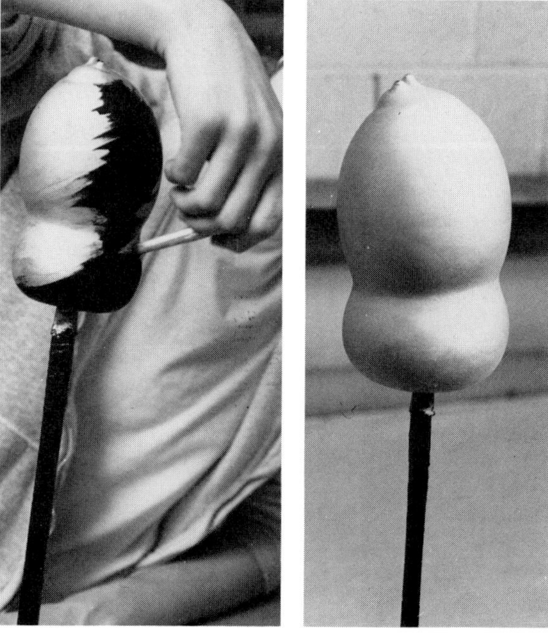

Once the matrix has been formed, it must be made conductive and attached to the copper cathode rod, which will provide electrical contact to the negative charge. The matrix is being painted with a silver conductive coating that will meet the cathode rod, making the whole electrically conductive. Note that if the form is to be hollow there must be an opening through which the matrix may be removed once plating has been completed. Wax may be readily boiled out through holes in the plating.

Far right: The wax matrix has been completely coated with conductive paint—except for two holes at the top. They will remain nonconductive, and will not plate, resulting in holes in the finished form.

The matrix is submerged in the plating tank and attached to the negatively charged bar at center. The tank pictured here is an industrial (300-gallon) acid copper plating unit in the metalworking and jewelry area of the University of Iowa Art Department. It measures 3' X 4' X 3'. In the background are pumps for filtering and agitating the solution. At each side of the tank hang positively charged copper anodes. These bars are gradually consumed during the plating process.

The plated form is removed from the tank. It usually takes from 8 to 36 hours to form a piece of this size. Thickness, which increases with time, is a function of need and the original detail of the matrix. The longer a piece is plated, the thicker the deposit becomes, and the more the original detail will be lost.

Several pieces await finishing. The first step in finishing is usually to remove the cathode rod and boil out the wax matrix in soapy water. The hole left by the rod may be covered by soldering a piece of sheet copper to the form. Otherwise, the pieces are finished by abrasive polishing and oxidation, just as would be any metal hollowware form.

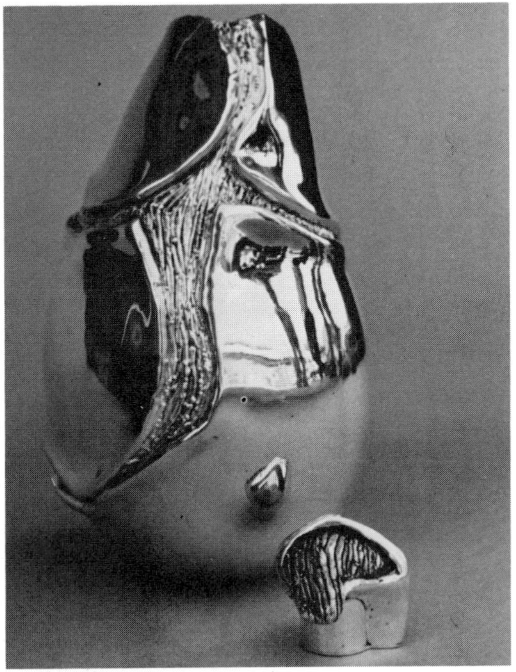

A finished electroformed container by Stephen Miller. *Photos by David Luck*

Barbara Anne Nilaūsen electroforms over a wax matrix as well. Each form and each lid are coated with conductive lacquer. Bodies are plated for 38 to 40 hours, while lids receive 12 hours of forming. The resulting thickness is approximately 14 to 16 gauge, although low density current areas are mainly 18 gauge. The electroforms are finished by filing, sanding, and scotch stoning until smooth where necessary. They are finally buffed with tripoli, white diamond, and then rouge compound. The surfaces are then cleaned and then readied for Nilaūsen's etching and epoxy inlay technique.

Below: Barbara Anne Nilaūsen often inlays her electroformed pieces with epoxy resin. The forms are first stopped out, except for those areas that are to be inlaid. In the acid bath that follows, the unprotected areas are etched. After 36 to 48 hours, approximately 1/32" of the electroformed surface will have been etched away, and the cavity will be ready for filling with epoxy resin and a variety of metal foils. After filling with epoxy resin, the material is allowed to cure, and is then filed, sanded, and polished so that the surface of the inlay is even with the surface of the electroform. An electroformed copper container with epoxy inlay, by Barbara Anne Nilaūsen. *Courtesy Barbara Anne Nilaūsen*

Below, right: "Canteen" (8" X 8" X 5"), by Daniella Kerner, is a container for very special beverages. Electroformed copper, nickel-lined interior, shoulder strap of vinyl tubing.

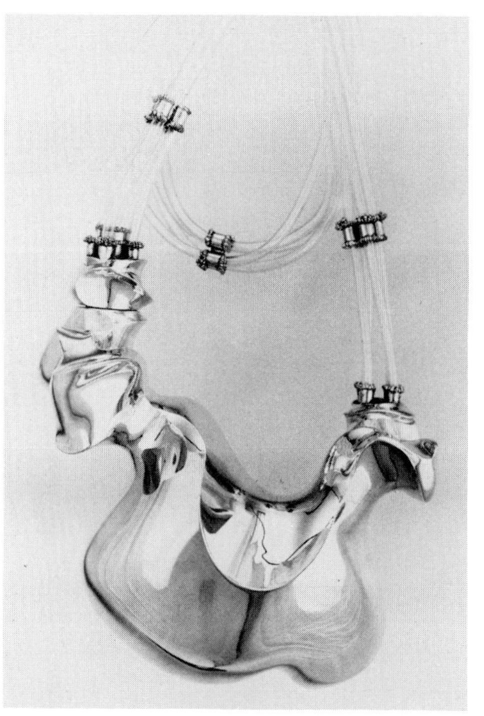

Silver Plating/Electroforming

To electroplate or electroform with silver, use the same type of tank setup as in copper plating. The only difference is the chemical composition of the electrolytic solution, and the fact that silver is required for the anodes. While silver is substantially more expensive than copper, it has the advantage of a unique luster when buffed, and for many craftsmen the look of silver is considered a premium. The chemicals needed for the electrolytic solution are silver cyanide, potassium cyanide, and potassium carbonate. The quantities needed are listed in the accompanying box. A word of caution is required. The odor of this solution is quite noticeable and noxious—extremely good ventilation is an absolute must. Also, this solution is extremely basic. It should be kept in a covered tank when not in use, since the addition of any type of acid to it can produce *extremely* hazardous fumes. Of course, never use the same tank for the silver plating that you use for copper plating; while the physical process is the same, the chemical action of the actual compounds is quite different.

Silver forming/plating can be effected on many metals, including copper, silver, and brass. Iron should not be introduced into the bath. Also, the Electrodag and other electroconductive paints are equally well suited for use in silver plating as in copper plating. Because the silver can be plated onto copper, many craftsmen—in order to save on silver expense—will have two tanks. First they will build up the bulk of their form with copper, and then plate a thin layer of silver over that copper electroform or electroplate. If you plan to use this combination, be certain to neutralize and clean the copper plated object before inserting it in the silver plating tank since sulfuric acid from the copper tank would contaminate the electrolytic solution.

Bright Dip for Copper

The silver can be polished using silver polish or buffing implements. In order to achieve an especially fine copper finish, or silver finish plated over copper, an optional procedure may be followed. By combining two parts sulfuric acid with one part nitric acid (Note: This is an extremely strong solution and must be stored carefully and handled only with gloves), you can create a "bright dip" solution for copper. When copper-plate or copper-electroforms are dipped in this brightener solution, the result is a *brilliant copper shine.* When dipping, allow the copper to be in the dip for only a split second. Then immediately rinse it in water. If you expose the piece for too long, the plated metal will be stripped away entirely. By electroplating for a while, then removing the copper piece, bright dipping, reelectroplating, and bright dipping repeatedly, a superior copper finish will be created. The smoother and shinier the copper finish, the brighter and smoother the silver finish will be after the copper piece is plated in the silver tank. Remember to thoroughly rinse the object and neutralize it before switching from *any* solution to any other solution. This bright dip cannot be used with silver-plated pieces; it can only be used to improve the silver surface indirectly, by improving the copper surface underneath.

9 CONTAINERS OF CLAY

The many forms of clay containers made today are the result of an aggregation of centuries of experimentation and experience. Immensely varied processes, simple and complex, create a rich body of information. It is no wonder that the contemporary ceramicist is in complete control of the medium.

Some potters rely upon that knowledge. They understand the underpinnings of the medium—the chemical and physical properties of clays and glazes. They design certain properties into clay by measuring and mixing components together. They compound their glazes, as well, to suit the firing properties of the compounded clays. These combinations are immensely varied and results often belie the degree of skill and control that the potter exerts.

At another level, clay workers can function very successfully without becoming too involved in the chemistry of ceramics. Clay shapes can be formed, decorated, and fired without time-consuming formulations. Much of the material for working with clay has been prepackaged—from clay to glaze. Formulas are keyed to proper kiln-firing temperatures with success almost assured. Modern ceramics, like baking a cake, can be a formula process with little sacrifice of aesthetics. To be certain, science and its new technologies, such as the electronic microscope, have cleared up ambiguities such as the size, shape, and character of clay particles. New insights have resulted in much greater control of clay, better clay, and a plethora of ceramic containers.

The range confounds: only a sampling is pictured here, yet the scope of processes is similar to what could have been found being used by clay workers at any point during the last thousand years. Slab- and coil-formed pots, wheel-thrown and mold-draped pieces are among the traditional and contemporary techniques. The basic properties of clay have long been understood, and its firing has been an art for centuries, producing porcelain, stoneware, raku-fired and reduction-fired pieces. Unless one can call thixotropic porcelain a new development, even the fine points of the craft are of ancient lineage.

The surface treatment of ceramic containers is time tried too. We still incise and carve, mold, model, allow the pattern to evidence the structure —such as coils showing where they join one another. And we still glaze and use slips in the same ways.

Rather, the innovation comes through the images of ceramic, the shapes and social content of the art. Sculptured elements have been added, areas are controlled and distorted in new ways. The ceramic container has become the vehicle of art. Sylvia Hyman's jars and boxes are political comments. Dave Boronda and Verne Funk use clay to poke fun at the human condition by utilizing parts of the human body in atypical and unconditioned applications. Other artists, such as Jan Axel and Greer Farris, stylistically express contemporary form, employing the idioms that they have invented.

Then there are containers that perpetuate the old traditions of ceramics, using classic shapes with beautiful proportions and glazes. These function in the wide range of ways they always have—as pitchers, bowls, and vases. (This does not mean that ceramic forms with social content cannot function effectively as containers. They certainly do in most cases.)

Susan Felix's hand-built container. Lid was formed in a mold. *Courtesy Susan Felix; photo by Sandy Solmon*

In the classic tradition, a porcelain vase by Jean Mann. *Courtesy Jean Mann; photo by Barbara Goodspeed*

Porcelain box with lusters in a contemporary style by Kathryn McBride (4" X 5"). *Courtesy Kathryn McBride; photo by Richard Barr*

Porcelain with clear glaze and lusters in an approach by Paula Winokur. *Courtesy Paula Winokur*

Hand-built container, "Hanged Man Floating in Limbo Land," by Elisa D'Arrigo. 9" high, cone 9 oxidation firing, glazed with china paint, lusters, and topped with feathers. *Courtesy Elisa D'Arrigo*

Judith Rothbart's teapot. *Courtesy Judith Rothbart*

Two boxes by Florence Cohen. *Courtesy Florence Cohen*

Verne Funk pokes fun at man and himself in "Come Fill My Cup." *Courtesy Verne Funk*

Another view of Verne Funk's chalice. *Courtesy Verne Funk*

A cookie jar, "The Seattle Tourist," by Joyce Moty. 13" X 10" X 18" high. *Courtesy Joyce Moty*

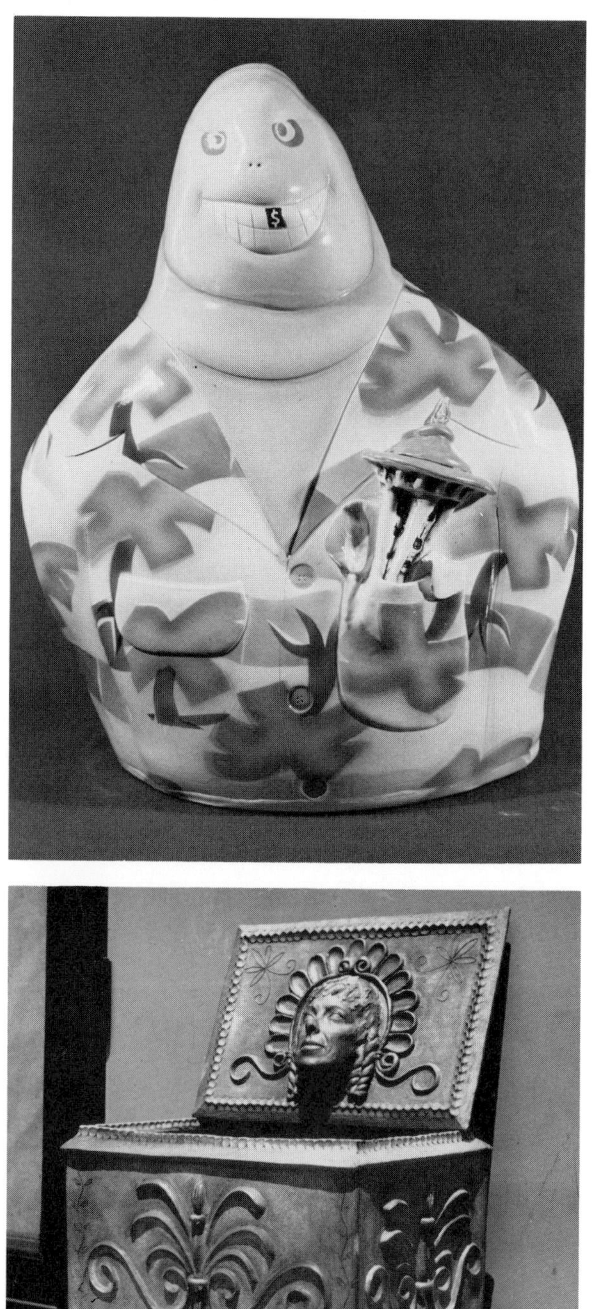

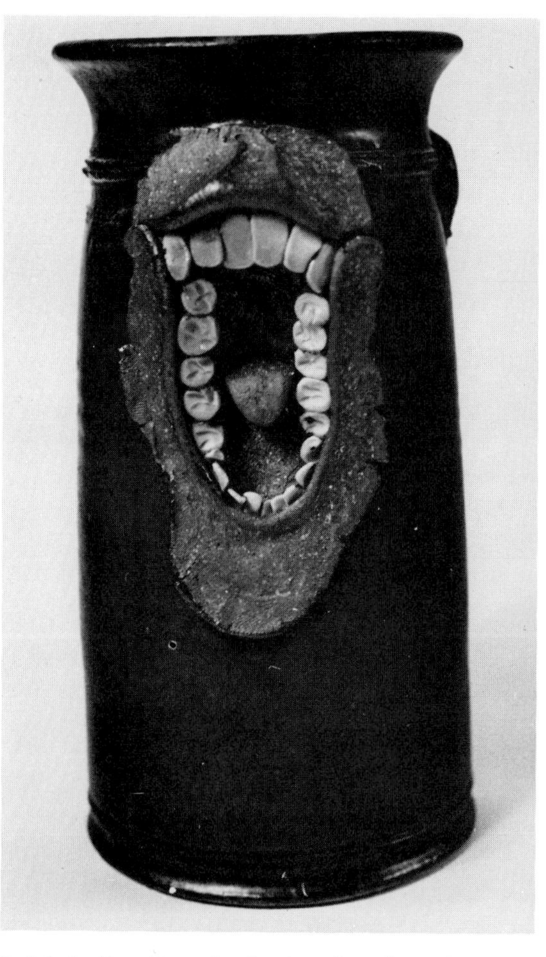

A stein by Dave Boronda. *Courtesy Dave Boronda*

"Sarcophagus for Myself" by Sylvia Hyman. Reduction-fired stoneware. *Courtesy Sylvia Hyman and the International Ceramics Collection of the Tennessee Arts Commission*

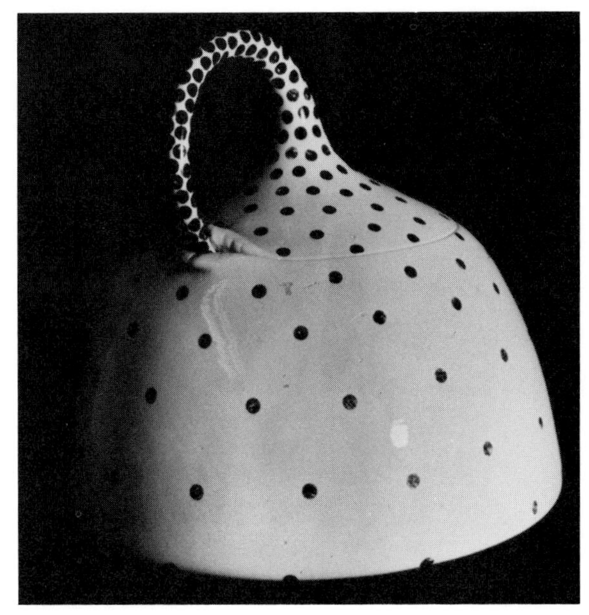

Verne Funk's "Dot Pot," decorated with his own decals. *Courtesy Verne Funk*

The Raw Material of Ceramics

After millions of years of weathering, feldspar (hydrated compound of alumina and silica) is worn down into microscopic particles. Water has combined chemically with each particle to form the basic element of clay—the platelets. When wet, these platelets slip and slide over one another, causing the mass of damp clay to stick together and to be manipulated into a shape. Clay is a thermosetting material; once hard, it keeps its shape. But before it has been baked to become permanently fixed, wet clay is plastic and can be readily shaped at room temperature.

Clays have varying degrees of plasticity. Ball clays are very fine grained and are highly formative. They often are added to kaolin, a less plastic variety, in order to produce porcelain. Most of the time, different clays are mixed together to create workable clay bodies with different degrees of plasticity and for specific cone-firing temperatures. For example, earthenware bodies are low firing; clay bodies for wheel throwing must be very plastic; stoneware bodies are usually prepared bodies that are very hard when fired; porcelain, also high firing, is a white clay that is a prepared body.

Ready prepared clays are ground, sieved, and blended. Sometimes they are sold in a powdered state. Moist clay must be stored in airtight, watertight containers, such as polyethylene garbage pails, so that moisture and workability are maintained.

Clay Preparation

Clay must be worked into a uniform texture that is free of air bubbles if it is not to distort or blow apart in the kiln. Kneading and wedging the clay accomplishes this. In the wedging process, clay is sliced and then slammed down on a plaster bat, one slice over the other, forcing out trapped air.

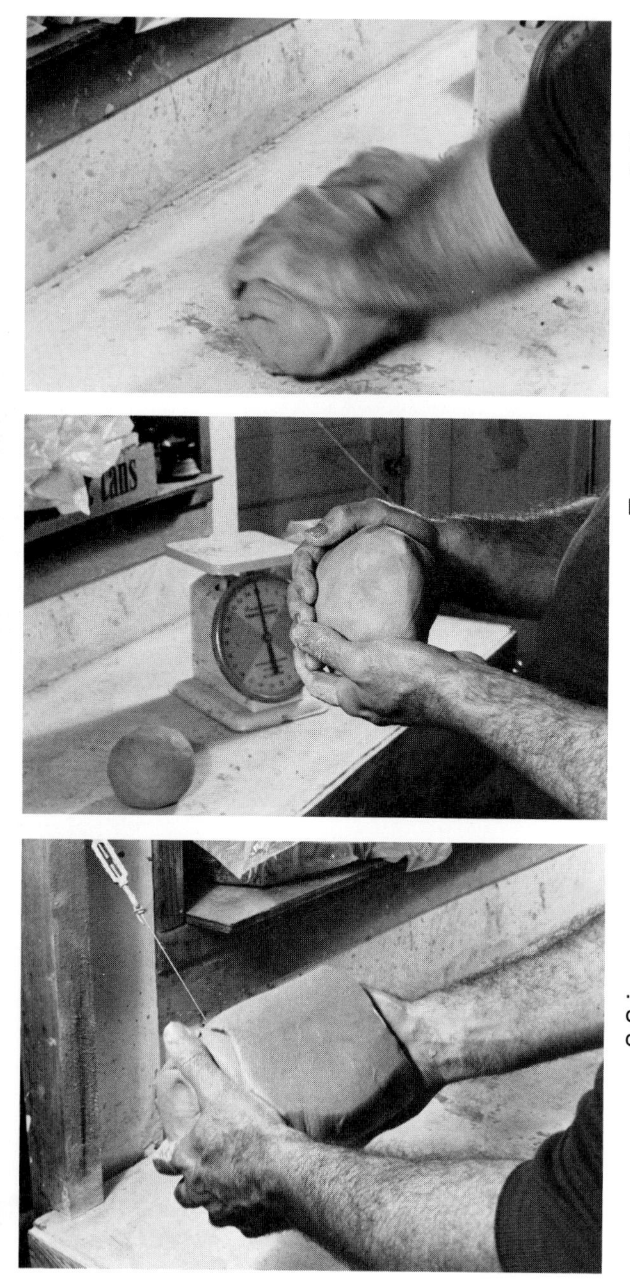

Paul Taylor prepares clay by kneading and wedging it on a plaster-topped table.

He compresses the clay into a ball . . .

. . . and slices it on a taut wire. Each slice is then slammed down on the plaster table, one slice over the other, to force out trapped air.

Container Forming

Pinch Pot

After clay has been conditioned it can be formed into containers in a variety of ways. One approach is to roll the clay into a ball and pinch a shape out of it by inserting one's thumbs into the center, rotating the pot and squeezing the walls up and out as the piece is revolved. The shape emerges gradually as the walls thin out. Pressure outward and upward determines the final contours.

THE CLAY QUALITY IN FORMING A STEIN by Dave Boronda

The clay is pinched and pulled and modeled into shape. Darker firing clays are added for the face and porcelain clays for the teeth and eyes. The form is trimmed and finished particularly at the base.

Left: A piece of clay is attached and pulled to form the handle of the stein.

Center: A thumb hold is attached to the handle to give the impression of being a part of the entire form.

Right and below: Two steins by Dave Boronda.

Photos courtesy Dave Boronda

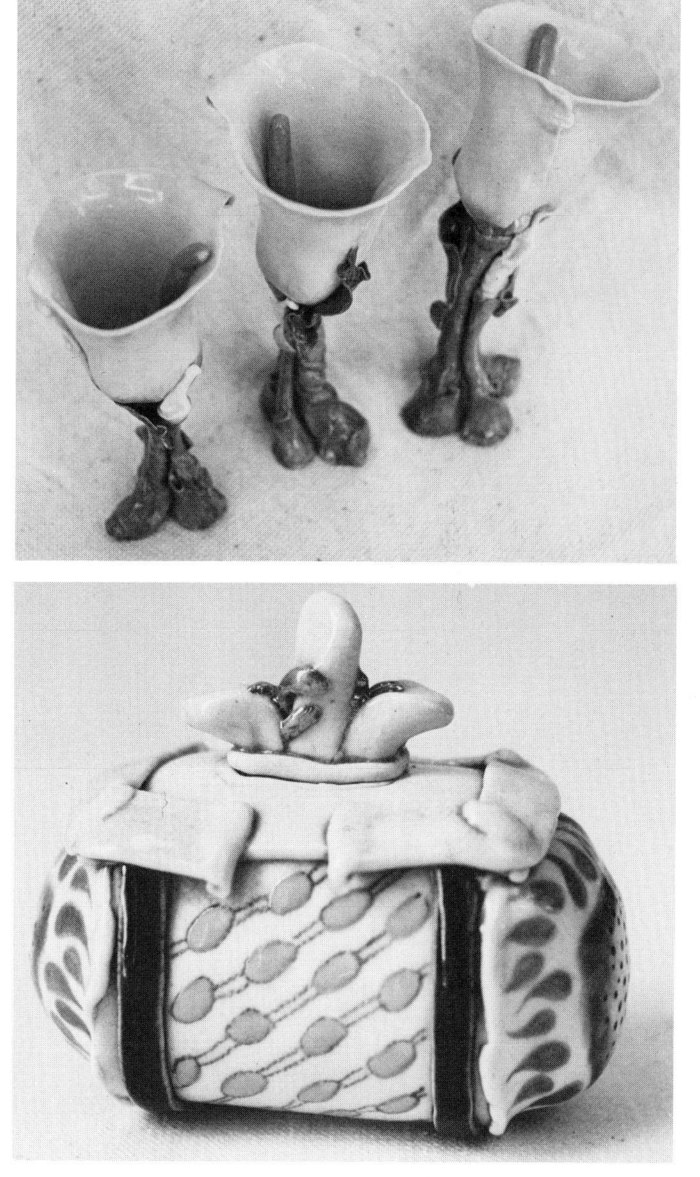

Clay can be pinched, flattened, patted, and shaped. Porcelain calla lily cups by Kathryn McBride (6" X 2"). *Courtesy Kathryn McBride; photo by Richard Barr*

A hand-built box by Kathryn McBride (4" X 2"). *Courtesy Kathryn McBride; photo by Richard Barr*

Coiled Form

Another approach is to roll pieces of clay into snakelike coils. Each coil is then spiraled around a center coil, edge touching edge, so that no air spaces are trapped. A pastelike clay can be used between coil edges. The shape can be controlled by decreasing or increasing the length of each coil. Upon completion, the coils can be flattened by means of a paddle, smoothed so that the definition of each coil is obliterated, or left in the "rough" state.

Clay can be rolled in snakelike coils and each coil built over the other. A box by Greer Farris. *Courtesy Greer Farris*

Slab Ware

Thick coils of clay can be rolled into flat slabs by using a rolling pin. The rolling pin rides on two sticks, one on each side of the clay. The thickness of these sticks will determine the eventual thickness of the slab. Shapes can then be cut from the uniformly thick slab of clay and these units can then be assembled. Joints must be properly sealed at corners and edges, fixing seams so that air does not become trapped. Trapped air expands in the hot kiln, exploding the form.

SLABWARE BOX by Susan Felix

After the clay has been rolled into flat slabs, it is placed on a mold. Pressure is applied with the same dowel stick (rolling pin).

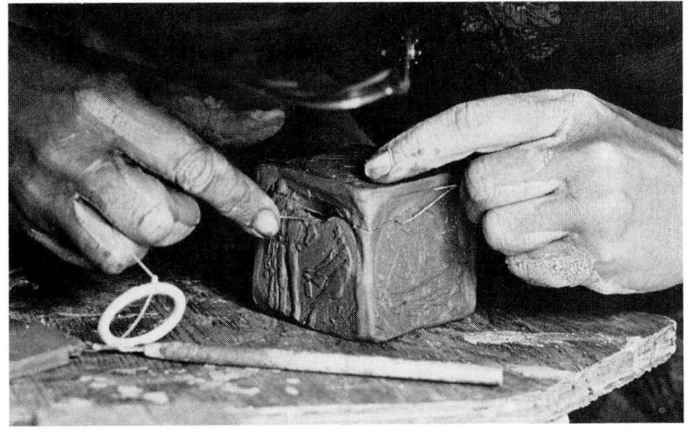

Sides are assembled so that a cube is formed. The lid is sliced off by means of a tautly held wire.

The box interior is refined, making certain no air bubbles are trapped in the joints.

A base is thrown on a wheel.

The base is then attached to the box. When dry, the box is fired and glazed.

A lining for the box, the same as the outside pattern, is block printed.

The completed box by Susan Felix.

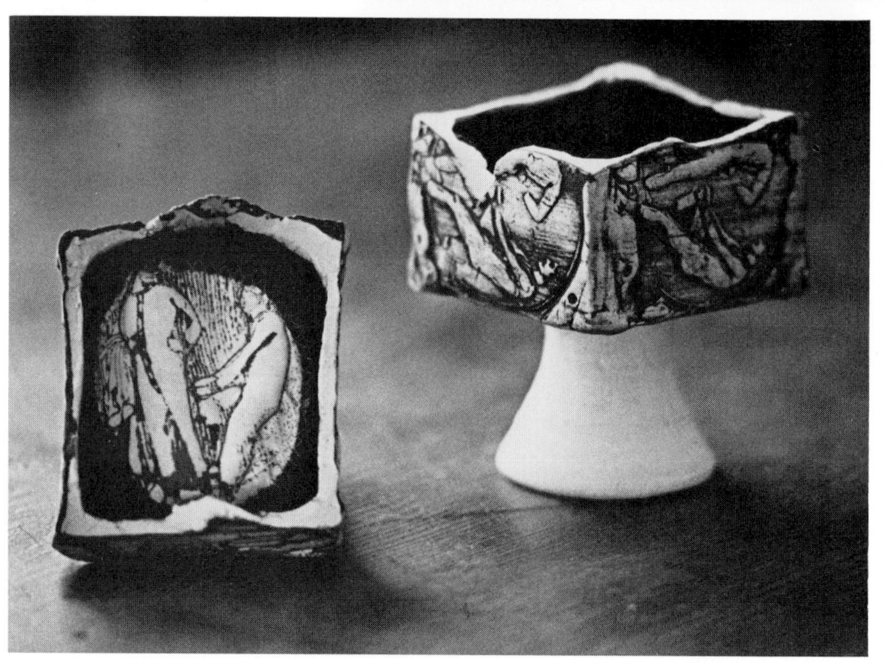

An open view. Note that the interior of the lid is showing the same figure (in cloth, though) as the design on the box sides. *Courtesy Susan Felix; photos by Sandy Solmon*

Slab face forms by Herb Schumacher. 7" X 4½" tall. *Courtesy Herb Schumacher, Chalk Creek Pottery*

Large slab vase by Albert Green. *Courtesy Albert Green*

Thrown Pieces

Clay can also be thrown on a potter's wheel. As the wheel revolves, centrifugal force causes the clay that is centered on it to move outward. This is probably the clay-forming skill that requires the most practice. In order to create a rounded form, all that is necessary is to increase the outside pressure; one hand is placed outside and the other inside the container, with the one inside slightly above the other. This forces the clay outward. Narrowing is accomplished by moving the inside pressure above the outside so that the final pressure is on the outside; or by collaring, which is applying pressure with both hands around the outside and no inside pressure.

THROWING A COVERED CONTAINER by Jan Axel

Prepared clay is secured to a plaster bat that in turn is attached to a wheel. The mound of clay is centered by raising and lowering the mound until it does not wobble.

Then the clay is flattened and compressed to the desired base size.

It is checked for the approximate diameter.

A well is made to open the pot.

The bottom thickness of the base is measured using an awl that is pierced into the base.

With the base the proper thickness, the bottom of the pot is extended to the outside wall.

The base is compressed and cleaned.

Walls are pulled up to the desired thickness.

The form is then shaped with a rib.

Excess clay is cut away with a taut wire or knife.

The top rim is smoothed using a soft, wet sponge.

The lid measurement is taken with calipers.

Excess clay is trimmed away from the bottom of the pot.

A pattern is pressed into the wall with a fingernail, leaving an indentation.

Now a clay lid is formed. A mound of clay is centered on the wheel and brought up to a cone. The top of the clay "hump" is opened for a lid and the correct inside curve of the lid is formed.

The lid size is taken to assure a match with the bowl opening.

The flange and eye are created and adjusted.

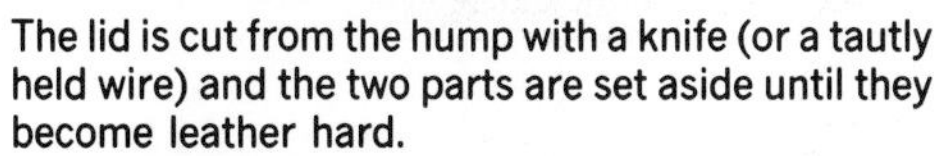

The lid is cut from the hump with a knife (or a tautly held wire) and the two parts are set aside until they become leather hard.

When leather hard, the lid is inserted on the jar base and trimmed so that the shape conforms to the opening and overall contours of the container.

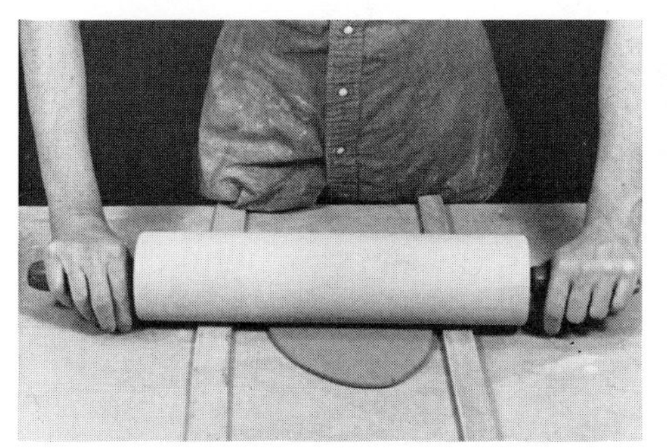

A slab is rolled out between two sticks (that determines thickness).

Circles are cut to the correct size using a cookie cutter.

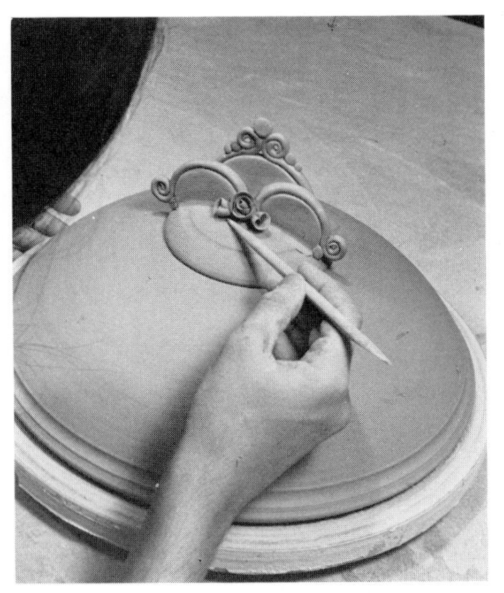

All decorations, such as slabs, coils, and rose forms, are attached securely with water and a wooden sculpture tool.

The completed piece is allowed to dry very slowly and when completely dry, the piece is bisque fired to cone 06.

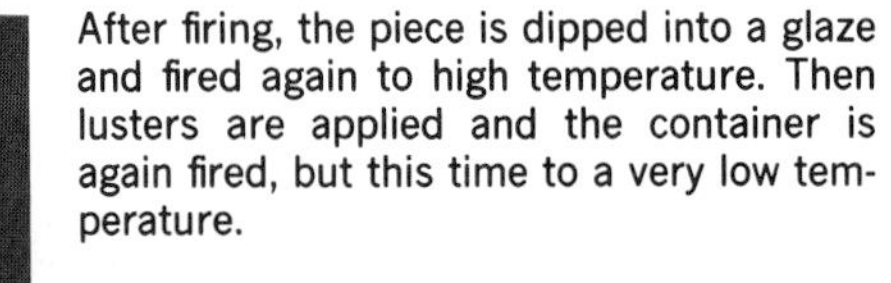

After firing, the piece is dipped into a glaze and fired again to high temperature. Then lusters are applied and the container is again fired, but this time to a very low temperature.

The completed jar in porcelain by Jan Axel (16" X 8"). *Courtesy Jan Axel; photos by Robert Axel*

THROWING A DOUBLE-WALLED FORM by Joel Moses

The clay is prepared, centered on the wheel, and opened. The piece is large and requires handling a large initial piece of clay. The walls are pulled up for the first form. The base should have a floor wide enough to accommodate the inside collar of clay that will be added. Calipers are used to estimate whether the proper size has been achieved. Good judgment has to be used as to whether the two pieces are in the proper proportion so that they will meet at the top.

After the outside bowl is finished, the interior base is prepared for introduction of the inside collar by scoring the floor. The area scored will coordinate with the piece to be inserted.

The collar is thrown immediately after completing the outside bowl so that the condition of the clay will be similar.

The cylinder is placed into the bowl and centered. Then the shape is further refined. It is very important to prepare the outside edge of the bowl before the two parts are joined by compressing and beveling the lip to an inward slope. This helps eliminate any buildup of excess slurry on the surface.

The inside collar is thinned out, flared at the top, and brought up to a greater height than the outside cylinder. The inside wall is extra thick to keep it from collapsing as it is flared out to meet the outer wall edge. The edge of the collar is also compressed and beveled so that the two beveled edges (inner and outer form) will match.

Beveling serves to reduce the collection of surface moisture and increases the area of contact. Before actually closing the forms, inside walls are sponged to remove excess water. The gap between the two walls is closed, using a flexible rubber or metal rib. Precision is necessary in order to facilitate the joining of both edges. (It is possible to flare the inner collar too much or too little.)

The seam is rubbed and smoothed. Then the entire outer surface is scraped with a rib to eliminate excess surface slip and moisture. The pieces are now in a "green" state, ready to dry. The trick is to effect an even drying for both walls to harden at the same time. If they dry and contract at different rates, the seam will separate. A small opening as large as a pinhole is made at some unobtrusive point to avoid explosion in the firing. (Heated air expands and, unless it can escape, the form will burst.) Firing is at stoneware with cone 10 glazes.

Joel Moses's double-walled pots. Fine white sand is used with a sandblasting gun (Model C, Sandy Jet Abrasive Blaster, A.L.C. Co., Inc., Medina, Ohio). The gun is powered by a compressor.

Photos courtesy Joel Moses

Teapot in stoneware by Robert M. Winokur. Salt glazed, blue wood ash, slips, and engobes make up the surface. 9¾" X 7¾". *Courtesy Robert M. Winokur*

Stoneware vessel with handle by Robert M. Winokur. 16¾" X 10". *Courtesy Robert M. Winokur*

Flowerpot by Albert Green. *Courtesy Albert Green*

Thrown cheese server in stoneware by Jim Cantrell. 11" X 9". *Courtesy Jim Cantrell; photo by Neal Cornett*

A lidded pot by Rose Cabot. Crystalline glaze in emerald green and charcoal brown. 9½" high. *Courtesy Rose Cabot, Tucson, Arizona*

Another lidded pot by Rose Cabot. Crystalline glaze in turquoise blue and olive. 8½" high. *Courtesy Rose Cabot, Tucson, Arizona*

Molding and Draping

Slabs of clay can also be pressed into a mold or draped over or into a form. The slab will conform to the new configuration. When leather hard, the new shape will be maintained.

MOLDED AND DRAPED BOWLS, Designed by Enzo Mari for Danese, Milano

Coils of porcelain are draped over a mold.

They are pressed (gently) to adhere them (and press out trapped air) at the point where two surfaces overlap. Edges are trimmed with a knife.

A mold is used to cut circles out of slabs of porcelain. Circles are overlapped and moistened at the point where they touch one another.

Model O designed by Enzo Mari. *Courtesy Danese, Milano; photos by Jacqueline Vodoz*

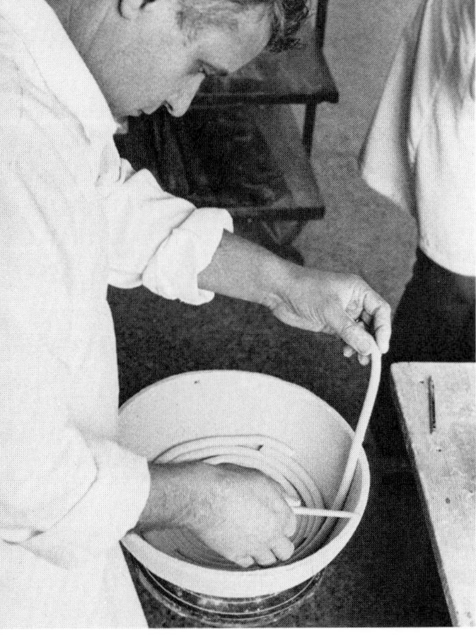

In another version coils are laid out in a plaster bowl-mold.

Model B, designed by Enzo Mari for Danese, Milano. *Courtesy Danese, Milano; photos by Jacqueline Vodoz*

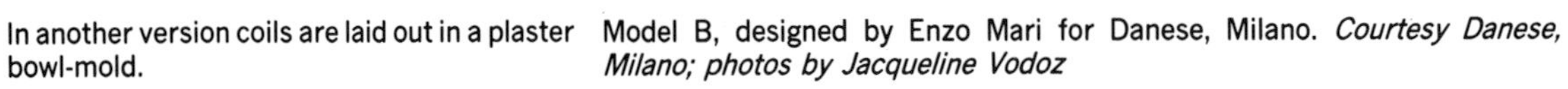

Composite of slab and wheel-thrown construction. Incised and applied clay decoration. Celadon glaze at neck of pot and on lid. Reduced to fire to cone 9. 7" X 6". By Angelo C. Garzio. *Courtesy Angelo C. Garzio*

Composite Pots

Lids, handles, spouts, feet, and appendages can be added using a combination of these processes. Also different processes can be used to shape various components of a single design. For example, the base of a pot can be thrown. The thrown piece can then be shaped by distorting part of it, and then coiling, in free form, is built onto the distorted thrown element to conform to the desired shape. The composite method invites a sculptural quality and permits the building of units into extremely large pieces.

A composite of thrown, hand-built, and sculptured elements by Sylvia Hyman. "Canopic Jars #1 and #2" in stoneware clay, 13½" high, 1970. *Courtesy Sylvia Hyman*

Decoration

After the basic form has been created, pieces can be individualized through various means of decorating.

Most containers will become complete statements just through the use of a suitable glaze. Others will call for other types of embellishment. Pieces can be press-molded by pressing a dab of clay into a mold and then superimposing it onto the form. Bits of clay can be carved out of or incised into the body of the container. Tools such as saw blades can be used to impart textures. Patterns can also be made by pressing a form (wood, wire, etc.) into the clay, leaving indentations. Slip, a water-diluted mixture of another clay color, can be painted onto the surface before the clay form has been fired. Sometimes scratches in the slip are used to reveal the undercolor in a process called sgraffito.

Glazes

Glazes are essentially glass. Glass is made of silica, which can be found in sand, flint, and quartz. Firing to about 1800°C. will melt these materials into a glaze. To raise or lower the melting point of silica in order for the silica to be compatible with a clay body, fluxes are added. Lead, soda ash, potash, lime, or borax are commonly employed fluxes.

In order to avoid stresses in the cooling and heating process and affect a permanent bond of the glass (glaze) to the clay body, some flint (glass) is added to a clay body and some clay body to the glaze. This helps to overcome compatibility problems. Therefore, glazes consist of silica, flux, clay, and coloring materials.

The principle of applying glaze to a bisquit piece (unglazed pot after the first firing) is to apply (by spraying, dipping, pouring, or brushing) nonsoluble materials in a water body so that the water is absorbed by the thirsty surface of the fired clay, leaving the unsoluble glaze material on the surface. In effect, the surface is sucking on the glaze coating. To make water-soluble materials into water-insoluble ones, the substance—soluble potash or borax and flint —is melted in a crucible and shattered into fine particles by pouring the hot melted "frit" into cold water.

Color is added to glaze with the addition of metal oxides (which do not burn away in firing). Oxygen in the air from water in the air combines in various proportions with metal oxides to form color. These colors vary greatly, depending upon the amount of oxygen permitted in the kiln (oxidizing atmosphere). If oxygen is excluded from the kiln altogether, as in reduction firing, the oxygen-hungry molecules, starving for oxygen, will seize oxygen from any source. Therefore, if a glaze contains iron oxide, the oxide is "stolen," producing a celadon green, whereas the usual final color of iron oxide in an oxidizing atmosphere is a tan or brown.

Glazes can vary considerably in texture from transparent glazes with glossy or matte finishes in various colors, to opaque glazes (addition of about 10 percent tin oxide) that produce a dense glaze. Introducing salt into a kiln atmosphere at about 1200°C. will cause the salt to vaporize, resulting in the production of hydrochloric acid and soda. These materials react with the chemicals (silica, flux, and oxides) in the clay body to produce a rough textured glaze.

If a glaze does not fit the clay body, then all kinds of surface aberrations occur. Crazing, crackle, peeling, and crawling are some of the effects. Sometimes they are considered desirable, other times not—depending upon the design, purpose, and statement.

Specific glaze formulas are plentiful in the books recommended in the bibliography.

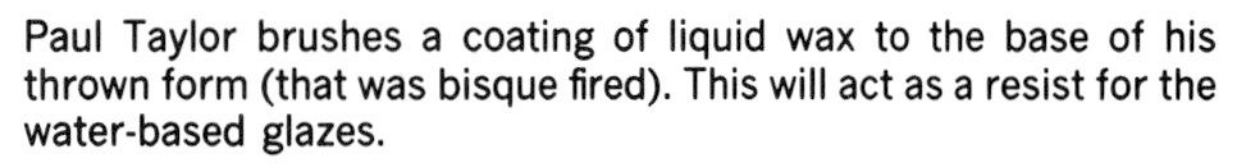

Paul Taylor brushes a coating of liquid wax to the base of his thrown form (that was bisque fired). This will act as a resist for the water-based glazes.

The pot is then dipped into the glaze and quickly removed. The "thirsty" pot quickly absorbs excess water.

The speckled glaze is poured into the pot . . . and immediately emptied back into the glaze mix.

Ceramic Process in Summary

Gathering clay: china clay, porcelain, terra-cotta, ball clay
Gathering additives to mix clay body: silica, grog (ground up fired clay), flux, clay, metal oxides
Optional: Purifying clay—removal of impurities by soaking, adding other ingredients, such as water
Conditioning clay: pounding, kneading, wedging
Shaping clay: pinch, coil, slab, draped or molded, thrown
Drying of clay container: first at room temperature and then in a warm spot
Container-fired in kiln: removing water driven off, chemical changes occur, surface becomes irreversibly hard
Result: bisque or bisquit pottery—container porous with unglazed surface. Terra-cotta or earthenware can be complete at this stage
Bisquitware glazed with water-suspended particles of glass by spraying, dipping, pouring, brushing
Glaze dries at room temperature
Container placed in kiln a second time—temperature buildup in kiln varies with clay body and type of glaze
Powdered glass ingredients melt and fuse to clay body
Form completed

Verne Funk's lidded pot was slip cast. The lips are added on. He makes decals by a photo silk-screen process using a dot screen with Kodalith film, Poly-blue film, or HyFylon screen material. Mason color and chemical overglaze ceramic inks are screened onto Advance decal paper and a cover coat of varnish is brushed over the whole piece. Decals are placed on a finished (glazed) piece and refired at a lower temperature. This pot was fired three times—bisque, glaze, and decal. *Courtesy Verne Funk*

A traditional stoneware glaze on a stoneware jug. The glaze effect, dripping and running of the color, relates to the function of the jug—to hold liquids. By Barbara Grygutis. 12" high. *Courtesy Barbara Grygutis*

Bottle with feldspathic glaze. 5½" high. By Herb Schumacher. *Courtesy Herb Schumacher, Chalk Creek Pottery*

Low fire luster glazes on stoneware by Florence Cohen. The effect is metallic, glossy, and highly colored.

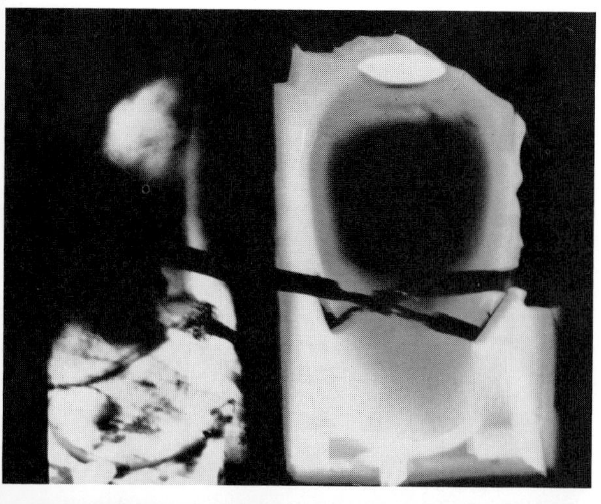

RAKU FIRING by Dick Luster

The raku kiln was improvised from a pile of broken M-23 Kaiser insulating brick augmented with swatches of Kaowool and asbestos. The burner was made from pipe fittings with a blower from an old clothes dryer. Propane was the fuel. The clay body was Van Howe Ceramic Supply #50—a talc raku body. The glaze is Soldner's transparent (80 Colemanite, 20 nephelene syenite) over engobes with various coloring oxides added to them. The engobes were applied at leather hard, and the glaze was added after a cone 08 bisque firing. The pieces were preheated in an electric kiln to about 400°F. and then placed in the raku kiln to heat slowly to about 1000°F.

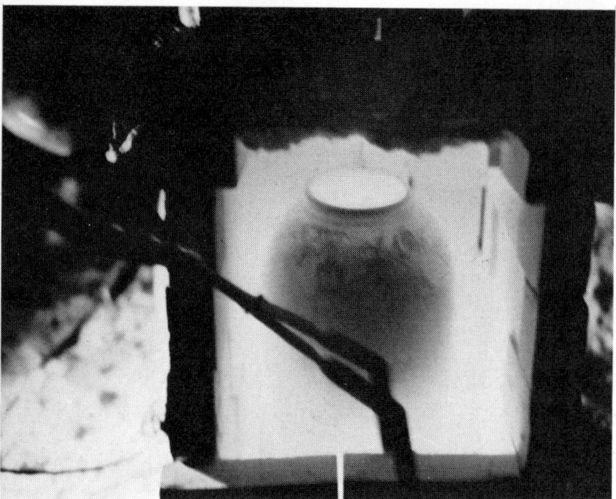

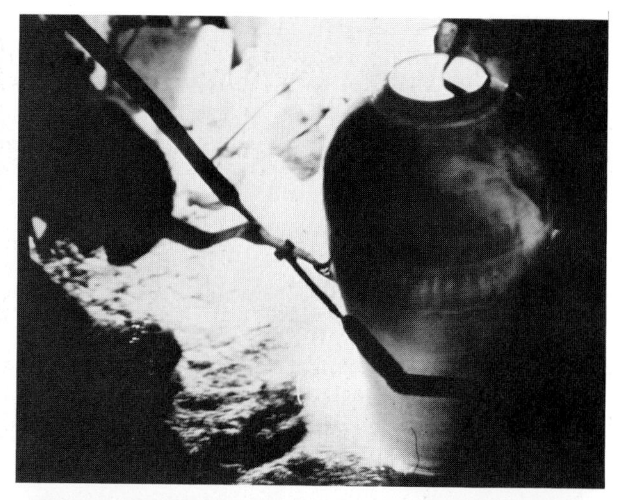

The pieces were then fired to about 1750°–1800°F. before removal from the kiln.

At that point the form was placed on a bed of sawdust.

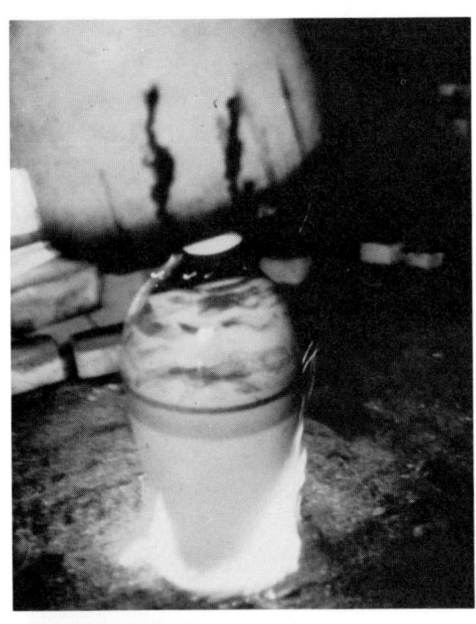

And a cylinder made of transite was placed over the form.

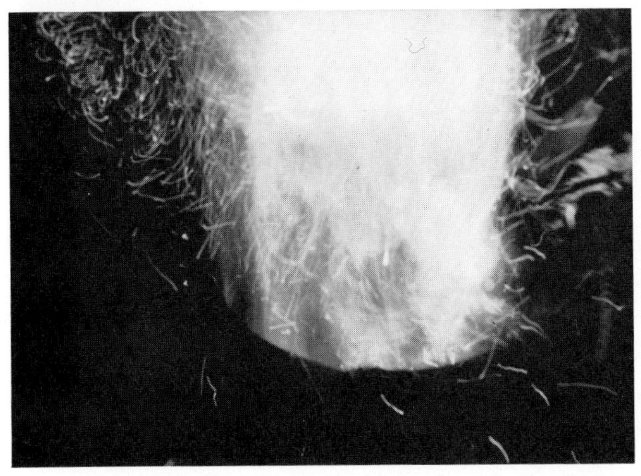

Sawdust was dumped into the cylinder (note the ignition of sawdust) and the cylinder was covered with a lid to smother the flames and contain the smoke caused by the igniting sawdust.

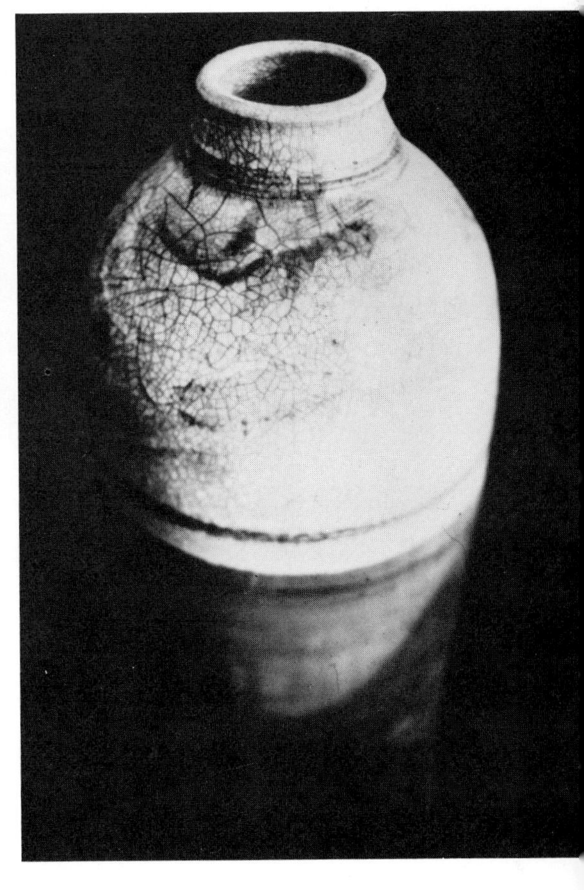

Large containers such as this one were permitted to cool in the covered cylinder for more than an hour before being allowed to air cool. (The temperature outside was nearly zero Fahrenheit at that point and when these pictures were taken.) Raku pot by Dick Luster.

Courtesy Dick Luster; photos by Tom Stephens

PAUL TAYLOR FITS A WOODEN LID FOR A POT

The mouth of the pot is measured for fitting of a lid.

A wooden lid is made by gluing together small blocks of wood and then turning the composite (inlay) on a wheel.

Here the form is sanded with 600 grit sandpaper.

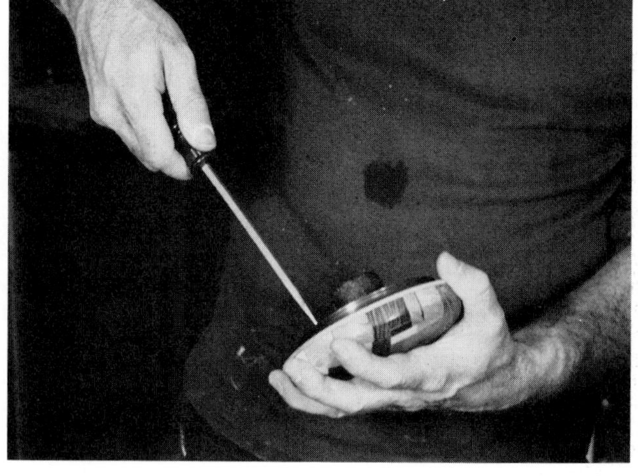

The faceplate is removed and the lid is ready for its cork mounting.

A round block of cork is measured and marked with a compass. This piece will plug into the container opening. It is cut on a band saw (and test-fitted in the pot). Then edges are sanded.

Two-part epoxy is metered directly on the area of cork to be adhered, mixed together thoroughly . . .

. . . and centered on the wood. (Brad nails temporarily indicate outer limits.)

Paul Taylor's jar.

10 STAINED GLASS CONTAINERS

The beginnings of the stained glass craft are lost in the blur of the distant past. We know that the windows of Saint Sophia at Constantinople were glazed, but whether or not the original glass was colored is questionable. Leaded stained glass probably did not emerge until several centuries after the advent of Christ.

Legend accounts for the discovery of gold-colored glass. The story is that a young Franciscan brother, who was working in the craft, was about to place some painted pieces of glass in the kiln when he was called from his work. As he rushed away, a silver button loosened from his clothes and dropped on one of the pieces. When the glass was fired, a lovely yellow stain had spread where the button once rested. That craftsman, Blessed James of Ulm, is the patron of glass craftsmen.

From clear and bright yellow glass, a complete range of richly colored glasses were developed. The good Abbot Suger of Saint Denis played an important part in shaping the development and interpretation of stained glass during the twelfth century. The cathedral at Chartres is a treasurehouse of twelfth- and thirteenth-century jeweled windows. Their full, brilliant color rivals the work of every age.

From the laborious process of painting each piece of glass to staining and coloring it, artisans developed other techniques. The working traditions continued but, during the nineteenth century, Louis Tiffany and his contemporaries learned how to make translucent glass, thereby eliminating the costly and time-consuming job of hand painting each piece.

Stained glass, of course, is best known for its use in windows and lamps, but contemporary container-makers have adopted the medium too. In fact, at a time when stained glass windows are becoming rarer, containers are beginning to be created in many shapes, sizes, and combinations of materials. Not only is stained glass readily employed to create square forms, but pieces can be combined into curves. As a container-making material, stained glass is just now coming into its own.

No matter what the form, all stained glass work exhibits a great consistency of goal: to display as much glass as possible with a minimum of connecting framework. It makes good sense, after all, to attempt to maximize color and minimize the interstitial elements that support the form. Working techniques have remained quite similar as well. The most traditional method of constructing stained glass forms—lead caming molded, cut to size, and soldered in place—is still used. But equally valid and often more flexible alternatives have also emerged. The most common and popular of these alternatives is the use of adhesive-backed copper tape. This tape was developed for, and is still used primarily by, electricians, but for stained glass craftsmen it has been the beginning of a new generation of forms. Essentially, the edges of pieces of glass are wrapped with copper tape, and the adjacent tapes on adjacent pieces are soldered to provide structural support to the form. This system provides great flexibility. Not only does it take much less time to wrap a piece than to cut lead caming and fit the glass, but copper tape makes it easy to use very small pieces and, most significantly, to create curves and shapes with great ease. To construct curves with traditional lead caming requires great skill and, even then, small curves are difficult. Of course, lead caming is still used by container-makers for very large forms where great strength is necessary, but most contemporary craftsmen have embraced the new method. It extends the range open to them without sacrificing aesthetics.

Making the Pattern

The first step in making any form is to plan it, and this is especially important in constructing stained glass containers. Containers are, by and large, not flat, and it is necessary to conceptualize the final form as a three-dimensional object. It is necessary to plan for any patterns that will move around the form, and it is necessary to plan for variations in the color and pattern of the glasses to be used. Glass must then be cut precisely to size. The best way to do this is by using a paper pattern that will yield patterns for individual pieces, which may be traced exactly in glass.

Depending upon the particular design, extra space between forms may need to be accounted for. The scissors used by glass craftsmen often have a special feature, a third blade that cuts a thin sliver off the edge of the paper pattern to accommodate the metal caming that will run between the pieces of glass. Trimming can accomplish this result too. Since that device was developed for use with traditional lead caming, it may not be necessary to worry about this where adhesive-backed copper tape is used.

Patterns should be saved, whenever possible, for use in later containers or in adaptations.

Types of Glass

Stained glass useful to container-makers is available in two forms—transparent and translucent. Within these two categories, an enormous range exists. Not only are there infinite varieties in color, but there are variations in grain, thickness, intensity, and purity as well. Every type serves its purpose in a particular application. Craftsmen generally plan not only a design, but also for qualities of the glass that will be used to execute the design. When selecting glass, observe the changes in coloration throughout the piece, and, when cutting specific forms, check to make certain that the color intensity combines well with the color intensity of bordering pieces. Observe variations in the grain. Will ups and downs look choppy or do they lend excitement?

Stained glass is available from specialists in every large city. Many are highly specialized, carrying the glass of only one factory or only one country. Use the telephone directory to locate sources. Another source is broken and discarded stained glass panels and lamps. If a piece has been damaged or discarded, there is nothing wrong with using the glass again.

Cutting Glass

Glass cutting is an essential skill. It requires practice and patience, but by following the proper procedures anyone should be able to master it after a short while.

The first step is to clean the glass. Use an ammonia-based liquid that will leave the glass free of grease. Then, lay the glass on a flat surface—wood, cardboard, or Neolite is best. Place the brown paper pattern on the glass, holding it with double-sided masking tape or Plasti-Tak, and cut directly around the pattern.

Glass cutters are available at most crafts and hardware stores. There are different types, but the primary difference is in the cutter wheel. Some are harder than others and will last longer—they also cost more. If held and used properly, all should cut glass precisely. Another variation is the ball end of the glass cutter. Some cutters have a heavy metal ball on the grip end that many craftsmen find useful in tapping along under the score line to initiate running the cut and aiding the break.

Most professionals grip the glass cutter, as illustrated, between the index and middle fingers and use the thumb to support the cutter from the underside. Hold the tool firmly, but do not tense your arm muscle. Place the cutter edge down on the glass until you hear the sound of the wheel cutting into the glass, then begin to draw the cutter along the pattern line, all the while applying just enough pressure to hear the wheel cutting into the surface. Applying the proper amount of pressure takes some practice. It is a good idea to experiment upon a few pieces of inexpensive or scrap glass. A properly scored cut should break readily—that is the test. A ragged line could produce a cut that will run off in an unwanted direction. Some people have difficulty holding the cutter between index and middle fingers and find alternate ways. The main point is that the cut has to be a solid, firm line.

Break each cut as it is made and never attempt to go over a score line; that damages the blade. Breaking the glass can be accomplished in several ways. To break a large piece of glass, lay it with score line parallel to the edge of a table—and slightly over it—and snap it apart. Or just snap it apart between your fingers of both hands. But if you are breaking long strips or small pieces, pliers may provide the best solution. *Running pliers* are wide pliers especially made for breaking long strips of glass; usually the larger part of the glass is gripped in one hand and the part to be broken away is gripped with the pliers. For smaller pieces, the same method may be employed using *grozing pliers* in one hand, or two pairs of grozing pliers. As before, experiment by cutting different shapes, different lines, and different curves. You will soon get the "feel" of how to score and how to break glass.

Although glass cutters will eventually get dull and worn out, with simple care most will remain serviceable through many applications. Occasionally dipping the cutter into kerosene before cutting will extend the life of a cutter. Store unused cutters in a small jar with flannel or steel wool at its bottom and add enough kerosene, or kerosene and a light oil, to keep the wheel and axle covered.

BASIC STAINED GLASS BOX

by Kay Wiener

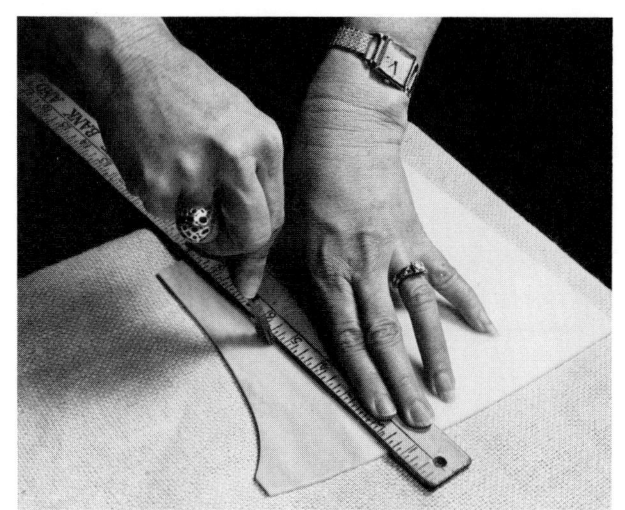

Kay Wiener measures and marks the glass, and cuts along a straight line using a ruler as a guide. It is essential that the glass be cut in a single uninterrupted stroke.

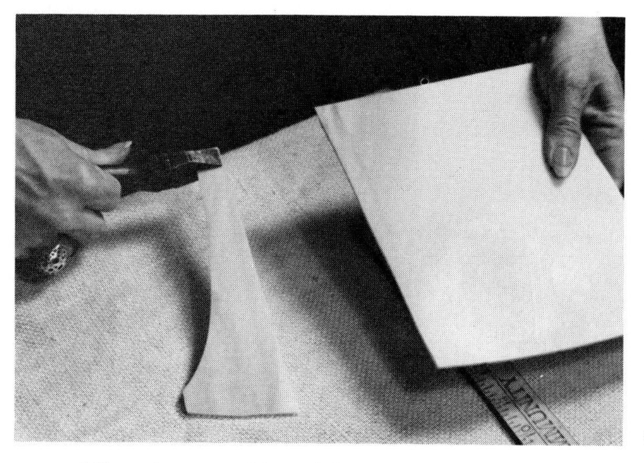

If the cut was a solid one, it is unnecessary to tap the glass from the back to make the score "run"; otherwise, a few gentle taps help. The glass is then broken along the score line. The edges are "grozed", or nibbled, with pliers to remove rough spots and sharpness.

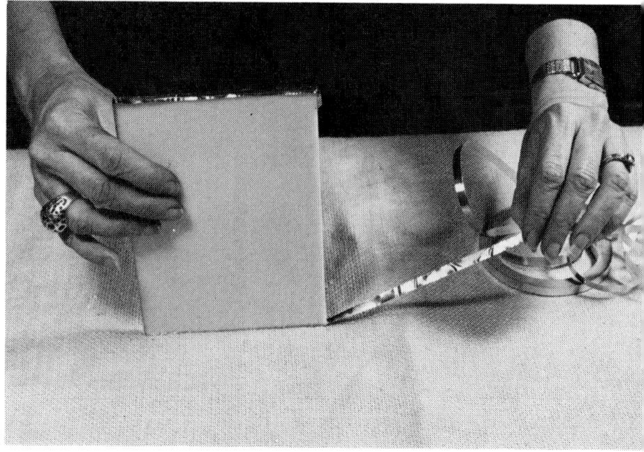

Adhesive-backed copper foil (available in several widths) is wrapped around the edges of the glass. The foil should be overlapped slightly where the ends meet.

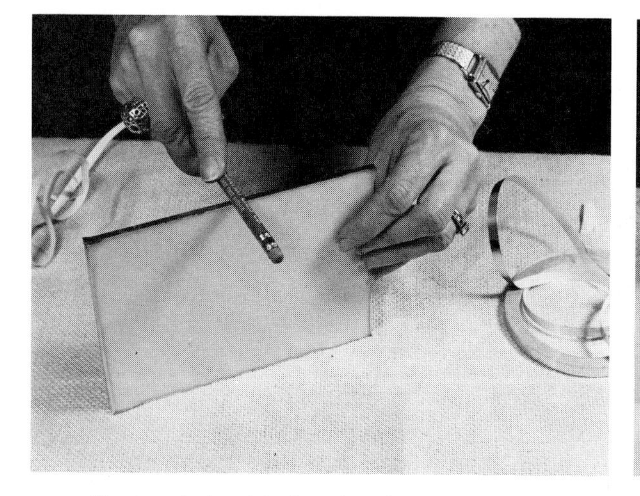

The tape is burnished to the edges of the glass with a pencil or dowel, and burnished to the sides as well.

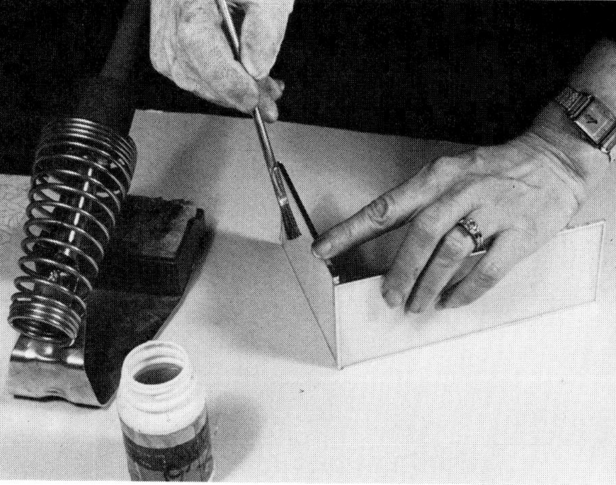

Once several or all the necessary pieces have been cut, grozed, taped, and burnished, they may be assembled for soldering. The pieces may be supported with masking tape or by hand. Before soldering, flux should be applied to the foil surface to dissolve oxides that inhibit the bond. The edges are then joined with 60/40 or 50/50 solder. The corners are first tacked to hold the forms in place. Then the full length of each side is joined to the abutting side. Once joined, all visible copper is coated with solder or "tinned," and a raised bead may be added for aesthetic reasons.

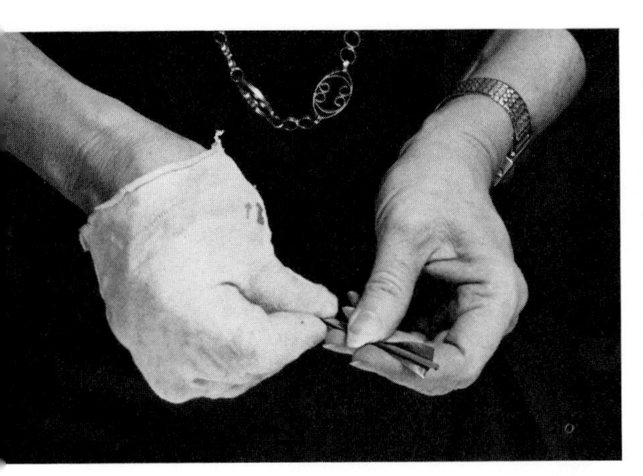

Commercial hinges may be used, but, here, Kay Wiener creates a hinge for the box top with 36-gauge sheet copper (not adhesive-backed foil) and a piece of brass wire. The copper was cut on a paper cutter to obtain an even ½" strip. It is then wrapped around the wire, which should project ¾" beyond each end of the hinge/box length. This wire will rotate within the copper foil.

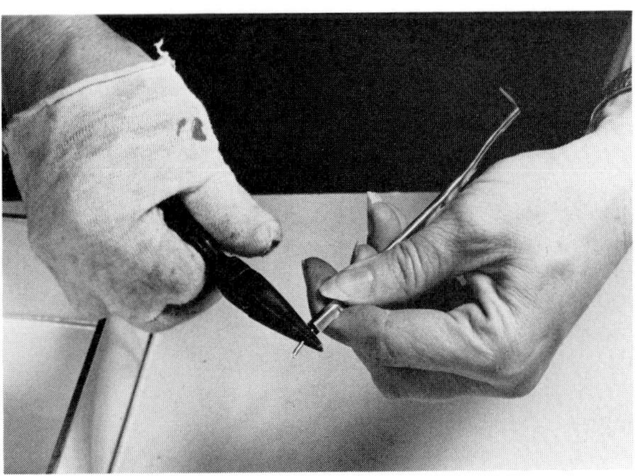

The hinge wire is bent back so that the ¾" extensions form a right angle to the copper tube. These extensions will be soldered to the back of the box.

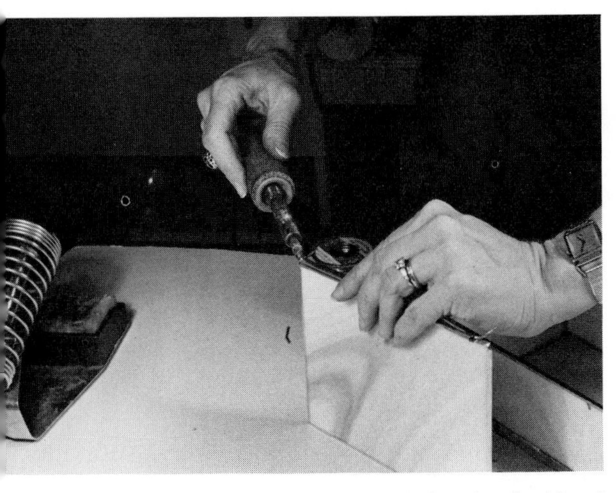

The copper tube encircling the wire is fluxed and soldered to the back of the box lid, making certain that the wire is not soldered to the top in any way. (In fact, it is sometimes best to leave the wire out until the copper tube has been soldered into place.)

The wire extensions are soldered to the back of the box, completing the hinge.

Kay Wiener completes the box by attaching brass findings.

"Habitat for Candles," of stained glass in copper foil construction, by Ruth Rickard. *Photo by Laposka Pix*

"Jeweled Box" (7¾" X 3½"). A geode slice is the central design element in this hexagonal box by Ruth Rickard. The container has a mirror bottom. *Photo by Laposka Pix*

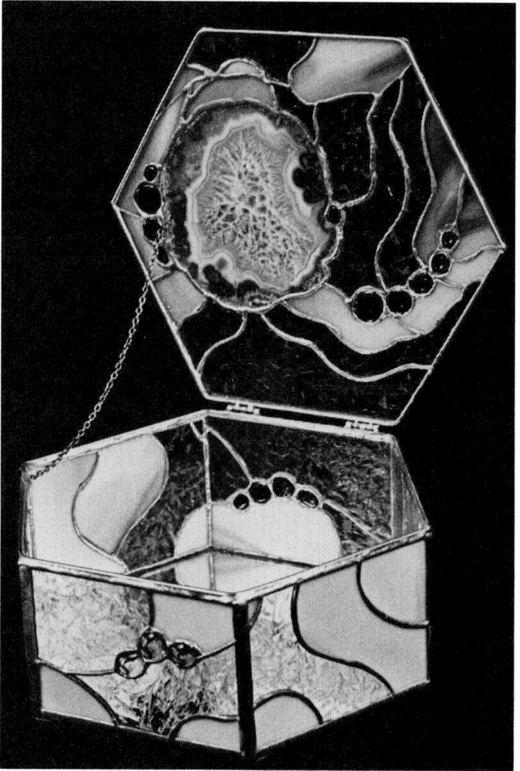

Malcolm Fatzer created this chest with doors and drawers, combining stained glass with mirror.

A clear glass box with variegated glass cabochon in the lid.

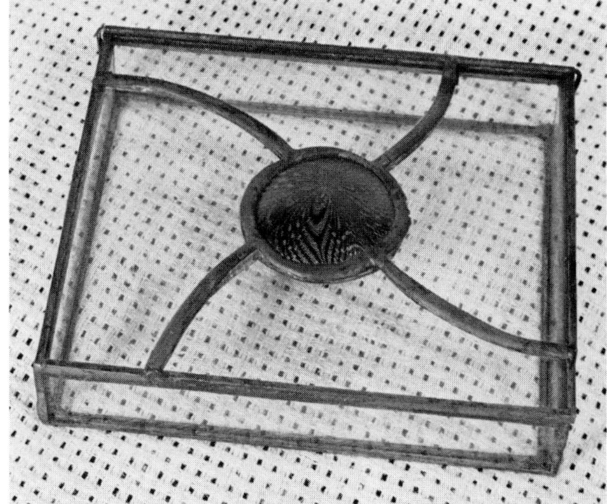

A rounded stained glass box by Frank Wright. *Courtesy Frank Wright*

←
Another box on legs—this one with a domed lid—by Carol Savid. *Courtesy Carol Savid*

A stained glass chest with curved lines and mirrored bottom by Carol Savid. *Courtesy Carol Savid*

Attachment Techniques and Construction

There are, essentially, two alternative attachment techniques. One uses the traditional lead caming, the other, the nontraditional adhesive-backed copper tape. Both are valid techniques. Lead caming is still used extensively in larger stained glass forms, but, increasingly, skilled craftsmen are employing adhesive-backed copper tapes on smaller forms. The reasons are clear. Although lead provides great strength, the copper tape is a faster, easier material to work with and it is lighter, a benefit on smaller forms. Actually, adhesive-backed copper tape allows craftsmen to achieve results that were either very difficult or impossible to obtain with traditional lead caming. Curved and shaped forms, difficult to construct with lead, are readily built with the newer technique. Of course, curves are possible with lead, too, but they usually require a larger diameter. Except for very large forms, copper tape should be considered a primary working technique, a method that professional craftsmen now employ extensively.

The technique of using thin channels of lead to hold intensely colored pieces of cut glass in place offers a perfect wedding of materials. Malleable, easily attachable lead fits readily around precisely cut pieces of glass. Though the lead is soft, the rigid material it encases allows craftsmen to create large, structurally sound units.

Lead caming, as these channels are called, comes in two configurations called "H" channel and "U" channel because the cross section of each is shaped like the letter. "H" channel is used between two pieces of glass with an edge fitting into the channel on each side. "U" channel—with only one groove—is used to finish edges, usually on the outside of the form.

In the traditional working techniques, "U" channel caming is laid along a row of nails that define the perimeter of the panel. Each piece of glass is individually cut and placed in position and cames are soldered into place to contain each unit, while nails hammered into the wood below hold the pieces tightly in place. Wherever possible, unbroken pieces of lead caming are continued throughout the panel, since continuous lengths create a stronger form than many short strips.

The preferred solder is known as 60/40, which stands for 60 percent tin and 40 percent lead. (Make certain that you use solid core, wire solder.) Some solders contain a flux core that only creates ugly oxides on the caming. Fluxes are necessary to prepare the surface of the lead to accept the solder, but most craftsmen find that a liquid flux—brands recommended for specific solders are available—works best. It should be brushed on immediately before soldering. Its function is to remove oxides that prevent chemical bonding of solder and metal.

Adhesive Copper Tape

A modern stained glass technique employs adhesive-backed copper electrical tape, as indicated earlier. (Scotch Brand tapes are highly recommended because the adhesive does not burn off when heat is applied during soldering.) One attribute of the copper tape is that it may be shaped after it has been applied to the glass. The copper tape is wrapped around the edges of each glass unit and is burnished flat. Each unit is soldered to its neighbor over a plastic, metal, or wooden mold. Soldering proceeds in precisely the same manner as with lead caming. The copper is first painted with a flux, and then solder is applied—first to "tack" the pieces together at a few points, and then to coat the tape entirely. The solder outline can be controlled by trimming the copper tape with a sharp knife before soldering.

Steps in Making a Stained Glass Container with Adhesive-Backed Copper Tape

The steps in constructing a container of stained glass with adhesive-backed copper tape as caming are:

1. Planning the container
2. Making a paper pattern and cutting out the patterns for each individual piece
3. Cutting the glass to shape with the aid of the paper pattern (as described above)
4. Grozing the edges of the glass to remove rough edges
5. Wrapping the edges of the glass with adhesive-backed copper electrical tape
6. Burnishing the foil onto the faces of the glass with a burnisher or another blunt tool
7. Excess copper tape on the face of the glass may be cut away with a sharp knife to create a finer line (optional)
8. Placing the pieces adjacent to each other so that they meet
9. Painting the surfaces to be soldered with flux
10. Tacking the pieces together at two or three points
11. Soldering them fully
12. Finishing the soldered areas
13. These basic steps are repeated until the entire form has been completed

STAINED GLASS CONTAINERS
BY Malcolm Fatzer

Right: Malcolm Fatzer marks his glass with a felt-tipped pen, cuts it guided by a T square, and breaks it by hand or with pliers.

Center, left: He cuts all pieces, grozes, tapes, and burnishes the tape, before assembling them.

Center right: The pieces of foiled glass are tacked, attached, tinned with 50/50 solder, and a raised bead of solder is applied. The bead cannot be applied easily unless the solder and foil surfaces are clean.

Excess flux is wiped off immediately with a clean rag.

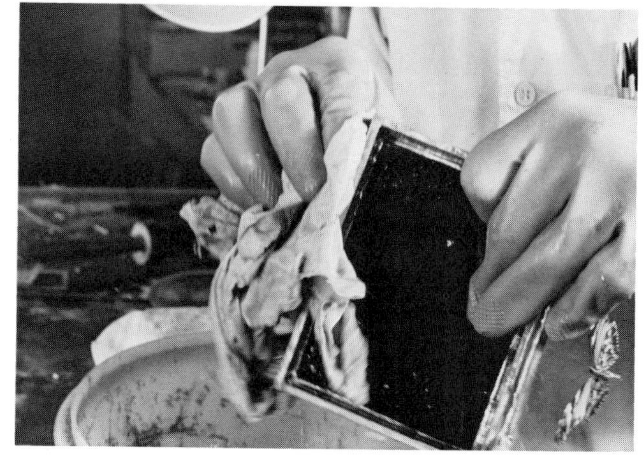

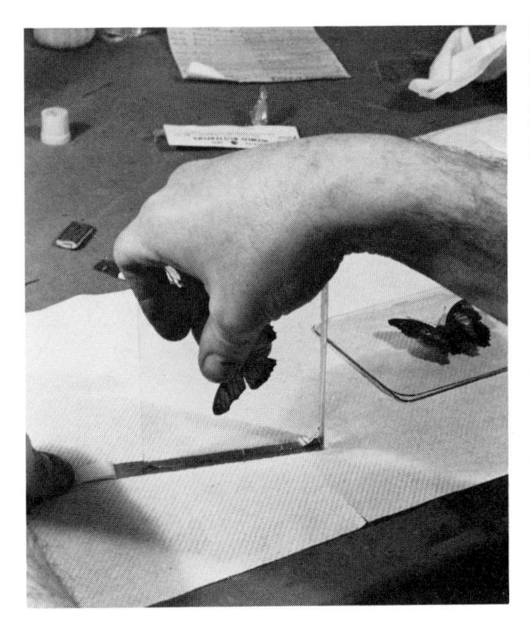

Malcolm Fatzer suspends a butterfly between two pieces of glass to create the lid for this box. He is extremely careful in cleaning the glass and in making certain that no dust, liquid, or solder seeps between the pieces at any point in the process. Here all work is done on a clean table covered with paper towels. Copper foil is carefully applied to the lid and burnished onto the edges and surface of the glass . . .

. . . using a shaped wood dowel to achieve a secure bond. Excess foil is trimmed away with a sharp knife. The final solder line can be shaped by shaping the foil.

Two hinges of ¾" long brass tubing are soldered to the back of the lid—always taking care to avoid penetration of the two layers of glass by any foreign substance. All soldering on the lid must be done quickly so that nothing seeps in to ruin it. Then all copper foil encircling the lid is tinned and beaded. Again, excess flux is immediately wiped away with a clean cloth.

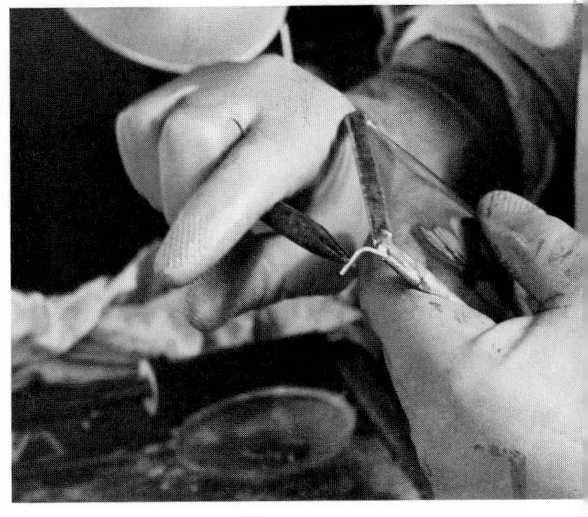

Brass wire is inserted into each hinge, bent back, and snipped off at the proper length. As above, this right-angled extension will be soldered to the back of the box. The solder is finished with a copper patina, achieved by rubbing it with a 40 percent solution of copper sulfate and water. Excess solution is removed with water. Again, the lid area must be kept clean and free of water. The edges are then polished with Wright's Silver Polish, wiped clean, and waxed with Johnson's Paste Wax.

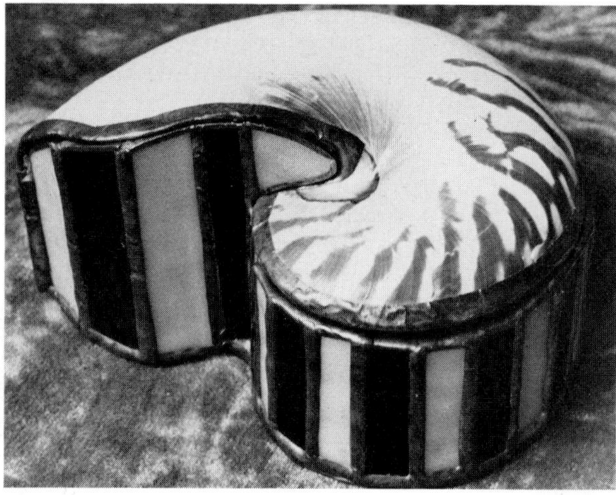

Above, left: A round stained glass container with dried flowers suspended between two pieces of glass in the lid. By Judith Anderson and Scott Bliss.

Above, right: Another stained glass container—this with a shell for a lid—by Malcolm Fatzer.

This stained glass container by Malcolm Fatzer includes a solder border around the shell lid.

NATURAL AND DECORATIVE CONTAINERS 11

As you can see, containers are made of almost every material and by as many different techniques, yet we have not considered one basic form. Throughout Africa and in parts of Asia and the Americas, containers grow on trees and vines and they are hard rind, inedible fruits—calabashes and gourds. Usually they are used as drinking cups, bowls, and boxes.

Another type of container adaptable from nature is a grass—bamboo. These tubelike growths have natural compartments in their culm that can be utilized as containers. The culm space (looks like the "trunk" of the bamboo plant) varies in diameter from parts of an inch to seven or eight inches, and the ridges along the culm may be a few inches to eighteen inches apart.

Then there are disposable bottles and boxes that can be recycled and transformed for another use by decorating their surfaces with decoupage, leather, shells, foil, modeling paste, and so on. Even a nondescript-looking wooden box can be enhanced by lacquering it or applying a straw mosaic. The decorative techniques shown here are just a sampling of a wider potential.

The Calabash or Gourd

Gourds or calabashes can be grown from seed and should be picked when the stems are very dry and ready to break. To obtain a dried skin, gourds are either heated or placed in the hot sun. The process is a slow one. The gourds should be wiped every day with a cloth that has been saturated with alcohol and turned so that another surface faces up. This drying process can take several months. When the seeds inside the gourd rattle, the fruit is ready for cutting and carving. A hole is first cut in the dried form to carve out a lid, usually at the stem end. Seeds shake loose and the form then is ready for decorating.

Another approach is to soak the calabash until the contents are completely rotten. During the soaking the skin becomes bloated and soft. Then the form is opened and the insides are cleaned out thoroughly. The emptied calabash is allowed to dry in the sun or in a warm room. The skin gradually dries into a hard shell. After the calabash hardens, decorating begins. The form may then be painted, carved, incised (lines scratched and filled with color), and burned with a wood-burning pen or a heated tool.

Procedure for Forming a Calabash or Gourd Container

1. Gather calabash or gourd
2. Prepare the calabash or gourd
3. Dry it until the skin forms into a hard shell
4. Clean it out
5. Decorate it
    a. carving
    b. incising lines
    c. burning in designs

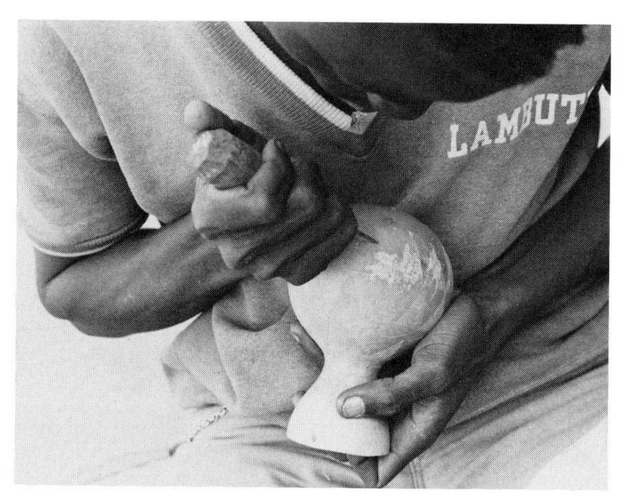

A Frafra boy of Bolgatanga (northern Ghana) uses an awl to scratch or incise lines in geometric patterns on the surface of a calabash (gourd).

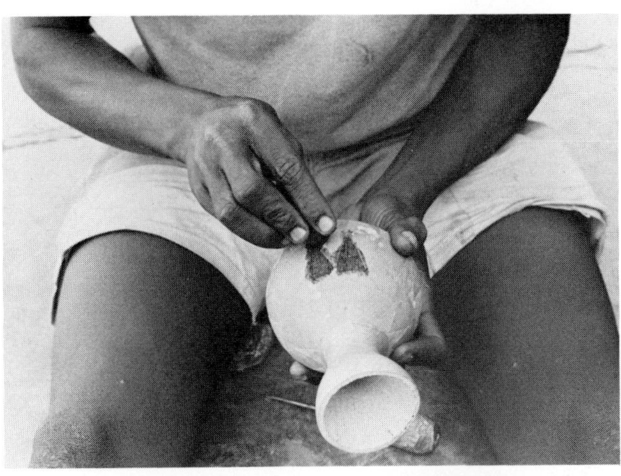

An oily (shee) nut is used for its brown black color.

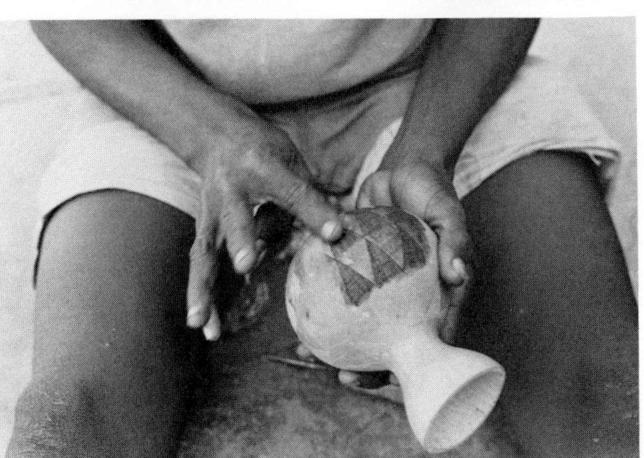

Excess is rubbed off with a finger.

Two calabash containers.

This large gourd (12" diam.) was incised in Peru using a technique similar to the African process.

The lid is cut in a puzzlelike configuration; the pattern determines the alignment.

An incised coconut bowl from Africa.

Gourds that grow on trees in Nicaragua.

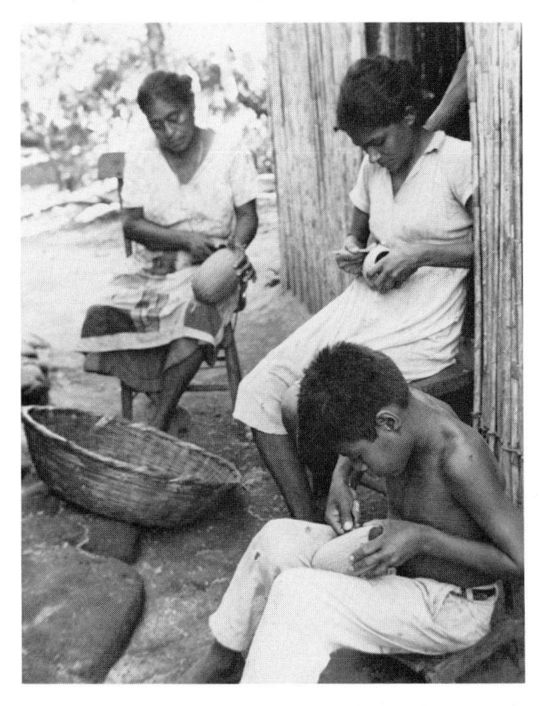

A Nicaraguan family carves designs into gourds

. . . using any kind of metal tool that will scratch lines into the woody shell.

The edge is trimmed with a knife.

A variety of gourds. The two on the left are from Africa.

Three gourd bowls carved by the Bush Negroes of Surinam, Dutch Guiana.

Designs burned into the surface of two gourds from Nigeria.

Two gourds from Mexico lined with soft beeswax with beads embedded in the wax. These are ceremonial bowls designed with fetish figures.

The Bamboo Form

Bamboo is an extraordinary plant. It flowers once in more or less a hundred years and then dies. It can complete its growing in two months and then remain the same size for the rest of its life. Most bamboo grows in clumps, grouped closely together. It thrives in the tropics, yet some species can survive freezing temperatures.

Harvesting can be programmed, usually in five- or six-year cycles. Age determines the qualities of the bamboo: The young plant is too soft for most craft purposes, while older plants are too hard and become prone to insect infestations.*

Decoration

The surface of bamboo lends itself to many decorative techniques. The hard outer layer may be scratched with sharp instruments and the spaces

*For those who are interested in growing bamboo, a beautiful book entitled *Bamboo* by Robert Austin, Dana Levy, and Koichiro Ueda (Tokyo and New York: Weatherhill, 1972) describes all aspects of the material.

A stock of bamboo in various widths, with different size culms.

filled with color. It may be stained and lacquered. Or designs may be burned into the surface with a wood-burning pen.

One of the most unusual and versatile techniques for decorating bamboo, is *lokub-a-lakusan,* from the Lake Maranao area of Mindanao, Philippines. Adapted for Western use, colored tissue papers (the kind that bleed when wet) are used to dye the surface of the bamboo. Tissue paper that has been cut into designs is placed around the bamboo and the entire form—paper and bamboo—is wrapped completely with thread. The piece is steamed or sprayed with warm water until thread and tissue paper are saturated and the color transfers to the bamboo's surface. Then the string and paper are removed to reveal the transferred design.

One of a kind *lokub-a-lakusan* from the Lake Maranao area of Mindanao, Philippines. Similar designs can be made using dye-bleeding tissue paper.

Procedure for Decorating Bamboo

1. Selecting proper piece of bamboo and cutting to size
2. Scraping off of outside covering
3. Decorating by:
    a. incising and filling in scratches with color or . . .
    b. carving away areas or . . .
    c. burning designs into the surface or . . .
    d. transferring colored tissue paper design
        (1) cut design in colored tissue paper
        (2) wrap tissue paper design around bamboo

(3) wind cotton string around paper to completely cover bamboo
(4) gently moisten with steam or warm water spray (do not allow water to run)
(5) remove string and paper

The Recycled Bottle

Wine bottles, jars, and other empty glass containers can be transformed in a number of ways. One way is simply to change the shape by cutting. Another process is to coat the clean bottle or jar with acrylic emulsion and, when that dries, to generously cover the glass form with acrylic modeling paste up to 1/4" thick, using a spatula. Once the paste surface partially hardens, patterns can be pressed into its skinlike surface and then lifted off. This impression remains and becomes permanent when the modeling paste hardens completely. One or two coats of acrylic gesso, or acrylic paint, applied over the modeling paste, completes the finish.

Another bottle-covering technique uses scraps of different leathers attached to the glass using an epoxy adhesive. Besides leather scraps, a wide variety of materials can be adhered to clean glass using epoxy cement. Shells, small pebbles, yarn, and cord are just a few of the possibilities.

An assortment of used containers waiting for recycling.

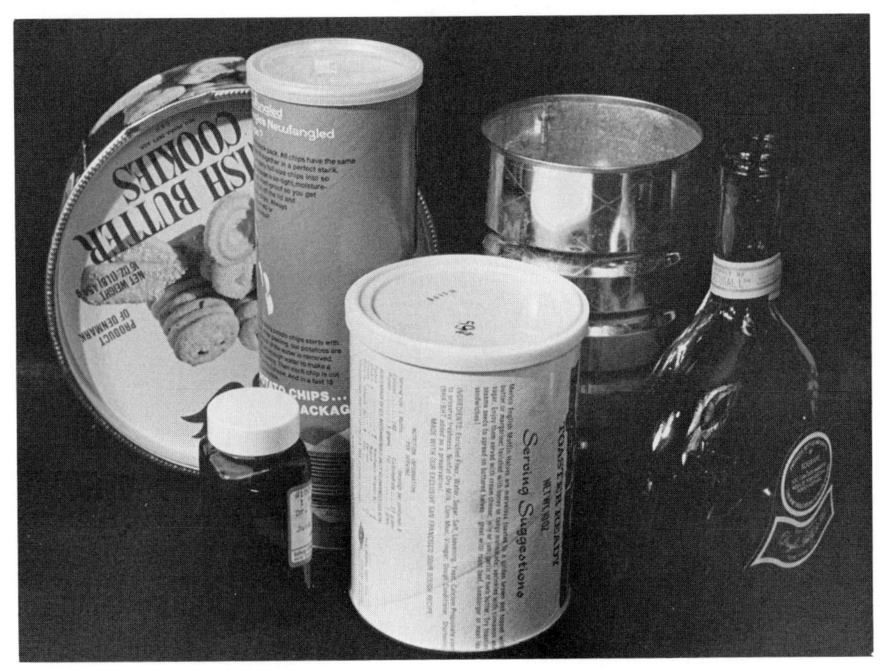

GLASS DECORATED WITH ACRYLIC GESSO AND MODELING PASTE

Left, top to bottom:
Clean the surface of the glass with alcohol, and coat the bottle with acrylic gesso.

Allow the gesso to dry, and apply a thick coating of acrylic modeling paste.

Allow the acrylic modeling paste to dry for several hours, until a skin forms on top, then press whatever is to be replicated into the surface. In this case, a piece of polyethylene doily will define the pattern.

Below: After the pattern has been made, allow the modeling paste to dry completely.

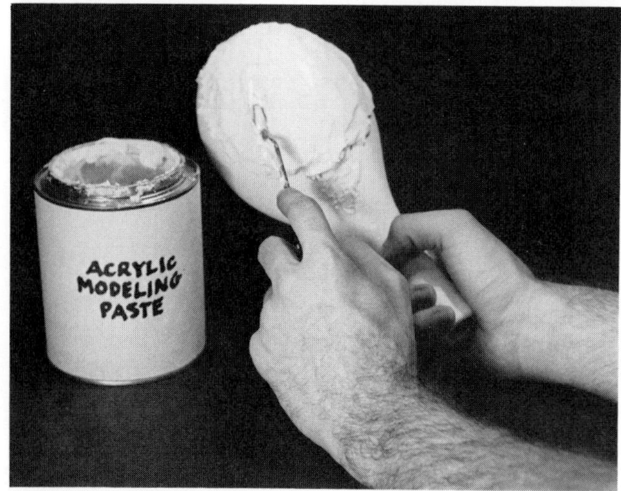

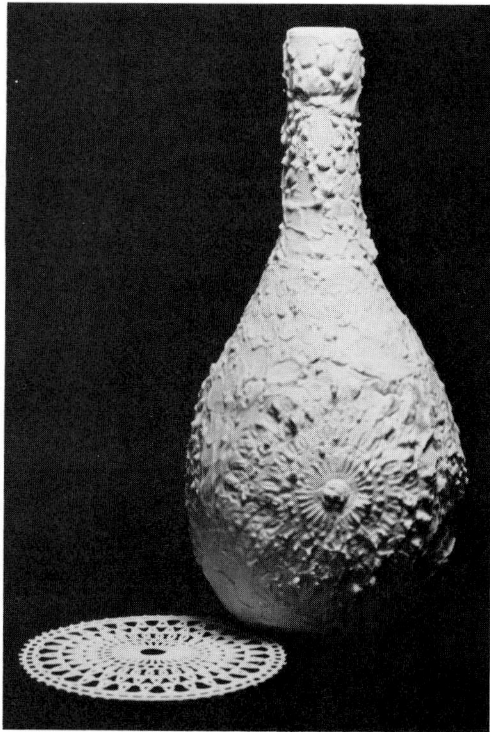

John Teeble covered an old silver chest with pieces of tortoise bamboo that came from an old roll-up blind.

He lined the interior with batik cloth.

Rush was used to cover this box, also by John Teeble. He painted it with orange shellac.

Scraps of leather were glued onto a bottle using contact cement. These bottles were decorated by Jeff Slaboden.

The Recycled Box

Any box surface, an ordinary wood (with little or no grain figure) or heavy cardboard, can be decorated. Some designs use split bamboo, rattan, and coverings of woven fibers. Others employ decoupage and utilize cutouts and varnish. Shells, straw, and metal foil repoussé offer other alternatives. And in the Persian box, a solid color painted background is decorated with lacquer; other techniques are inlaid lacquer and a lacquer scratch process.

Applying Decorative Elements

Prior to any decorating operations, the surface of the basic container must be prepared, either by cleaning or by painting with a background color prior to attaching surface elements. Wooden boxes should be sanded and the wood sealed with a sealer. If made of metal, grease and dirt should be removed with alcohol.

Shells, bamboo, rattan, and straw can be adhered using a two-part epoxy or with Sobo, Elmer's, or other white glues.

DECORATING A WOOD BOX WITH SHELLS

Left: An unpainted wooden box is stained with a black alcohol stain. A five-minute two-part epoxy is mixed and spread on the outside of the box.

Bottom, left: Shells are sorted and then arranged in mosaic fashion on the epoxied box.

Below: When the epoxy has cured, a very durable bond results.

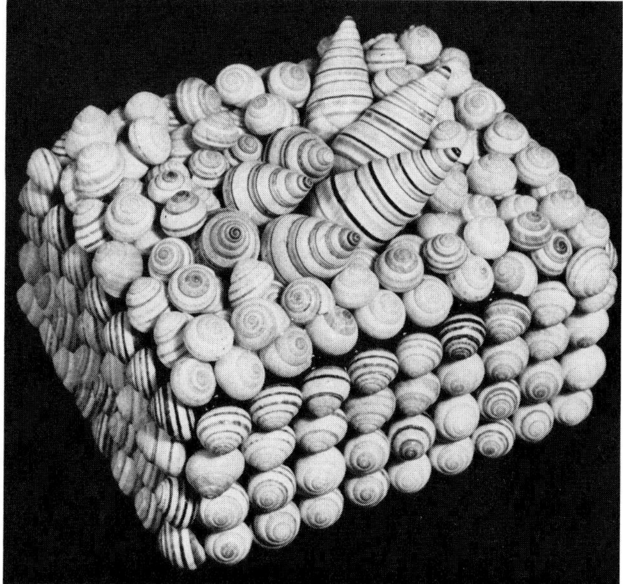

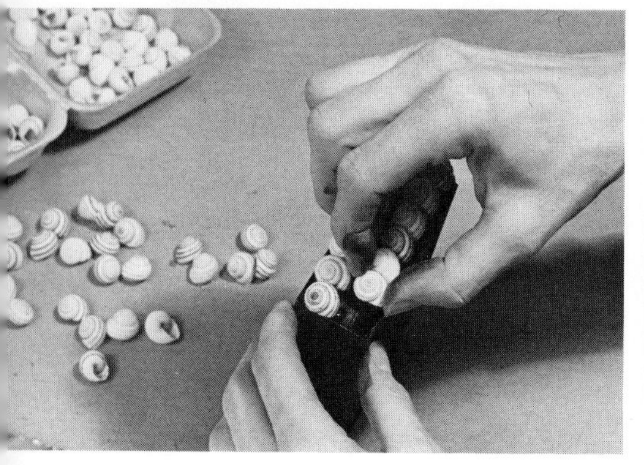

COVERING A TIN CONTAINER

An assortment of woven grasses.

A woven palm place mat is cut for use as a surface covering for a tin box. Contact (rubber) cement is sprayed onto the box.

A piece of the mat is glued to the surface of the metal.

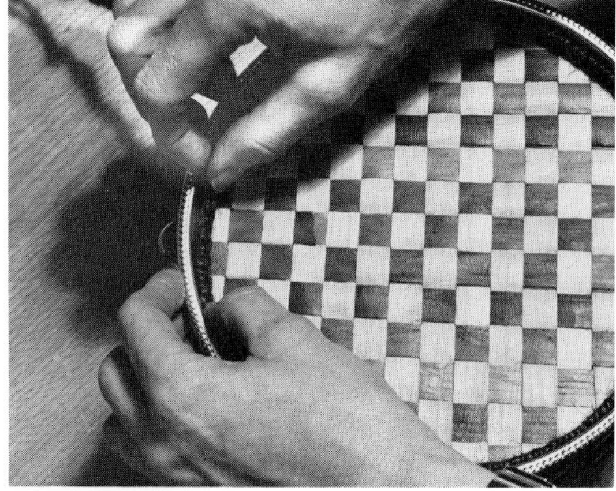

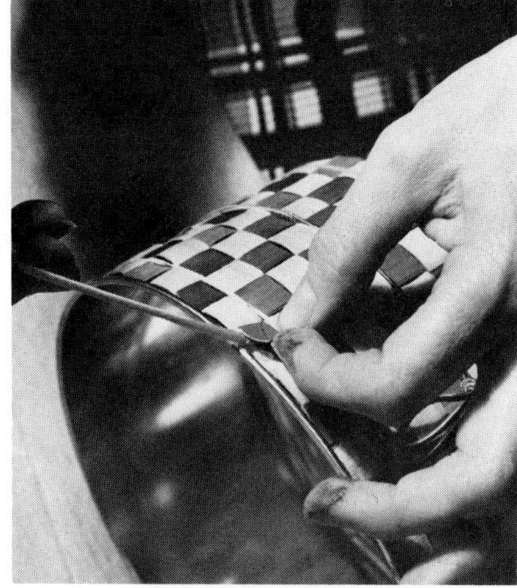

Top: More epoxy is applied around the edge . . . and a piece of braided yarn is attached to trim the edge.

Center: Similarly, palm mat is glued around the box. Where woven edges stick out, unattached, they are glued down to prevent unraveling.

"The completed box."

Decoupage

In decoupaging, finely cut paper or fabric elements are glued to the surface with a white glue that has been diluted slightly with water. For fabrics, specific white glues like Velverette are best. When the glue-covered paper is pressed onto the base surface, all air bubbles must be removed. After the glue has dried, many coats of varnish (from 5 to 25) are brushed over the entire surface. Each application is allowed to dry before applying the next, and after every few coats, the varnish is sanded with successively finer grits of wet or dry sandpaper. All the grit is wiped away before applying more varnish. When a sufficient number of coats are on, perhaps enough to bury the paper in the buildup of coatings, the piece is more finely sanded (up to 500 grit) and then waxed.

Procedure for Applying Decorative Elements

1. Prepare the surface that is to receive the decoration: clean, sand, seal, stain, paint, etc.
2. Plan your design
3. Attach the elements with the appropriate adhesive
4. For decoupage:
   a. Coat the entire area with many layers of decoupage varnish
   b. Sand smooth with successively finer grits of wet or dry sandpaper
   c. Wax

Below: Dee Davis's hand-colored decoupage box in an eighteenth-century design.

Right: A Scalamandré silk lining color-keyed to the exterior.

A matchbook box in decoupage by Alice Balterman. The black background between mosaiclike units acts as a unifying grid. *Courtesy Alice Balterman*

A desk basket in decoupage by Alice Balterman. *Courtesy Alice Balterman*

Wallpaper is used to trim the sides of a wooden tissue box. The trim is decoupage braiding. Elmer's glue was used to adhere all materials. Ten coats of varnish completed the piece.

Foil Repoussé

Repoussé is the process of raising or embossing a flat metal surface until parts appear in relief against flat areas. In order to do this, some nonrigid surface is placed beneath the metal. A thick pad of newspaper or felt serves well. Then sticks or metal tools (there are many sizes and shapes and curves) are pressed into the back of the metal to create indentations in the desired pattern or design. The degree of relief depends upon the amount of pressure and whether the indentation was burnished by repeated pressure, rotation, and vertical or horizontal movements of the tool. Sometimes, to further define the raised area, reverse pressure is applied around the repousse on the face or front of the metal. Of course, heavier gauge foils require more persistent pressure and repeated movement of the tool. Lighter metals can be indented very easily—even with a fingernail. If a heavy gauge metal is used, other techniques are employed as described in chapter 8.

When adhesive-backed foil is used, as shown here, repousse is accomplished while the protective paper covering the adhesive is in place. This is removed after the pattern has been raised.

Foil can also be repousséd in a pattern by placing the foil over some textured surface and then burnishing or rubbing it with a tool. If the foil is thin enough, the pattern will come through—either in a positive or negative raised image (depending on the pattern being "rubbed").

Procedure for Metal Foil Repoussé

1. Select object to be covered and clean surface
2. Decide on design
3. Press design into foil
4. Apply adhesive-backed foil to object with pressure, or apply a contact cement to foil and object surfaces. When tacky, attach foil
5. Touch up with tool by applying pressure on all indented areas

COVERING A CONTAINER WITH FOIL TAPE REPOUSSÉ

Arno foil, an adhesive-backed aluminum tape, is modeled into a design, while the piece rests on a soft surface.

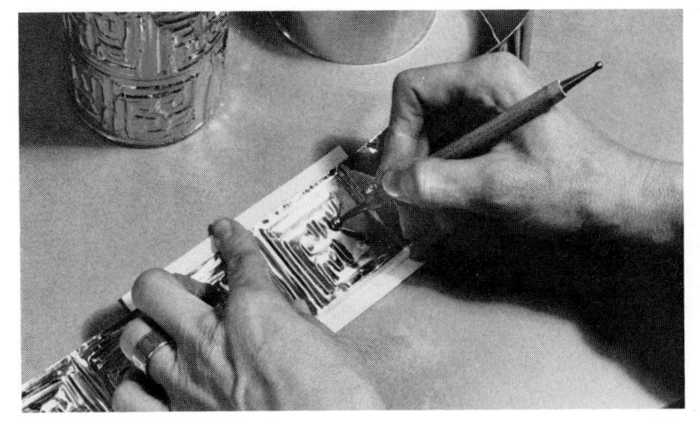

A protective backing is peeled away, and the adhesive-backed foil is pressed around the container.

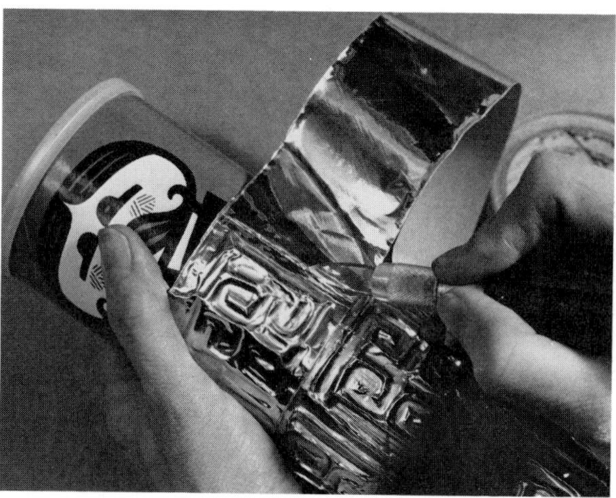

Excess foil is trimmed with a knife so that the blade slices through both layers, effecting an exact match.

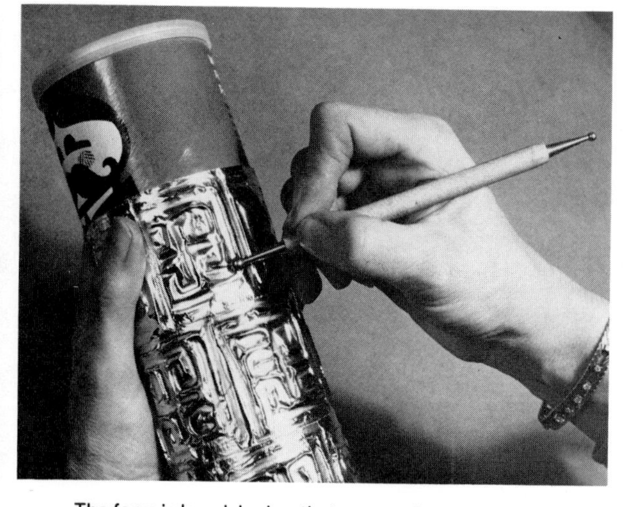

The form is burnished so that recessed areas are firmly fixed to the surface of the container. Where the ends of foil meet, the seam line is burnished so that the juncture is nearly invisible.

The completed container.

Arno foil tape used to cover a box. *Courtesy Arno Foil Tape*

More box designs with Arno foil tape. The cover of the box in the foreground is woven with strips of tape. The one in the background was rubbed over a doily and then antiqued with black acrylic paint. *Courtesy Arno Foil Tape*

Split Straw Mosaic

Straw crafting is an old art, an outgrowth of the harvest when straw from grains such as wheat, rye, barley, and oats was used for its stem (nothing went to waste). Besides weaving straw into baskets, the stems were used to create mosaic designs on boxes.

Each grass stalk consists of the head—the grain or plant's seed, a stem, and a thin husk that covers the stem, attached at nodes. When the grain is removed and the husk is pulled away (husk can be used too), a hollow straw tube remains. (This was the original drinking straw.)

Straw should be gathered in early summer when it ripens and turns a golden color. Rain, or any prolonged exposure to water, tends to discolor it. The stalks should be cut near the ground and laid in the sun or in another warm place to dry.

When dry, the grain is removed and the husks stripped off the stem. The straw may be dyed, if desired, using aniline or natural colors in either warm- or hot-water dye applications.

The material is prepared for mosaics by soaking straw in hot water overnight. Straw is less brittle after soaking and can be split and cut into shapes using a sharp knife, razor, or small sharp scissors. Cut pieces may be flattened by pressing with a warm steam iron.

Begin by arranging the small elements into a pattern. (Small pieces can be arranged on masking tape.) Then attach them, one by one, to the box surface with a touch of white glue (Elmer's, Sobo, etc.). You can use a needle to aid in the process of transferring tiny bits from one surface to another. Allow the glued straw to dry thoroughly before applying a protective coating of clear lacquer, varnish, or acrylic emulsion to the entire surface.

Procedure for Creating a Straw Mosaic

1. Gather and dry straw
2. Prepare box: sand, seal, coat
3. Dye straw if necessary and soak it overnight in warm water
4. Cut straw with a sharp instrument into desired shapes
5. Press flat with warm steam iron, if necessary
6. Attach to box surface in a predetermined pattern using white glue
7. Allow glue to dry
8. Apply protective coating to entire surface

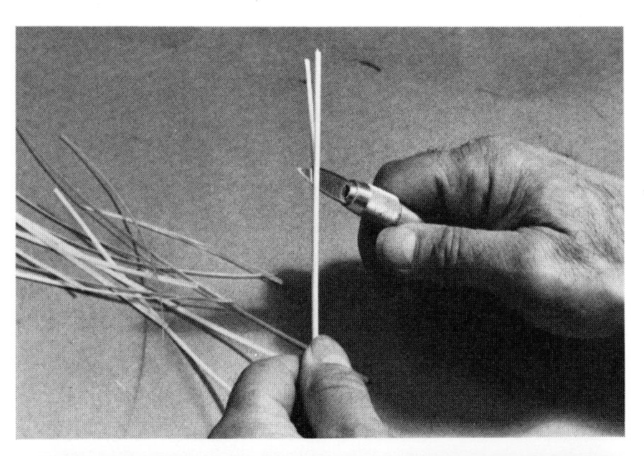

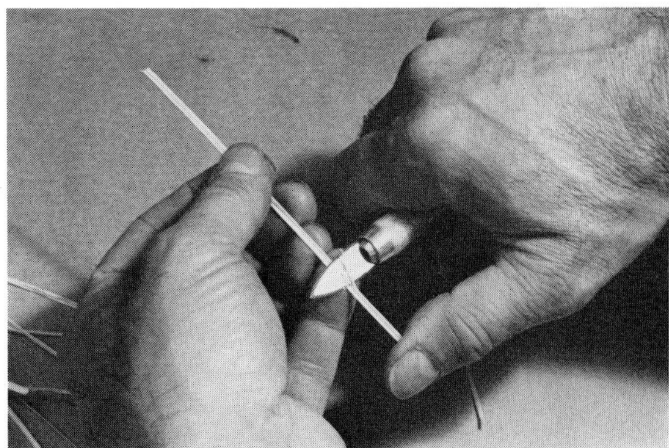

DECORATING A CONTAINER SURFACE WITH GRASS

Top, left: Dried grass is stripped of its outer sheath and split longitudinally with a knife.

Top, right: Curved edges are pared away.

Center: Parts that require dyeing are dyed using natural (vegetable) colors. When dry, pieces are cut and glued with white glue to the container surface. Box from Russia.

Bottom, left: A Russian box with a snowflake design in straw.

Bottom, right: Anna Continos's "Escollage" of natural plant materials, pressed and adhered with white glue to a box. After gluing, the entire piece is coated with a diluted acrylic matte polymer.

Lacquer as Decorative Design

Traditionally, lacquer has been a colorful and durable surface covering for wood. Its ease of application has made it just that much more practical and appealing.

Three basic applications are shown here. One is painting a picture (or design) with colored lacquers onto a wood surface. The other two are variations of lacquer processes popular in Mexico.

In one, the laca rayada process, colored lacquer is brushed over the surface. When thoroughly dry, a thick top coat of lacquer filled with clay is applied over the base coat. When the top coat partially dries—enough not to be tacky but soft enough to be scratched away—lacquered areas in the top coat are scratched away to reveal the first layer of lacquer. A feather fitted with a thorn is the tool used to remove the lacquer. After scratching in the design, it takes about eight days for this lacquer to harden completely. Before hardening, mistakes can be removed by burnishing the still soft lacquer with a highly polished stone—to spread it and cover the error.

LACQUER CONTAINERS

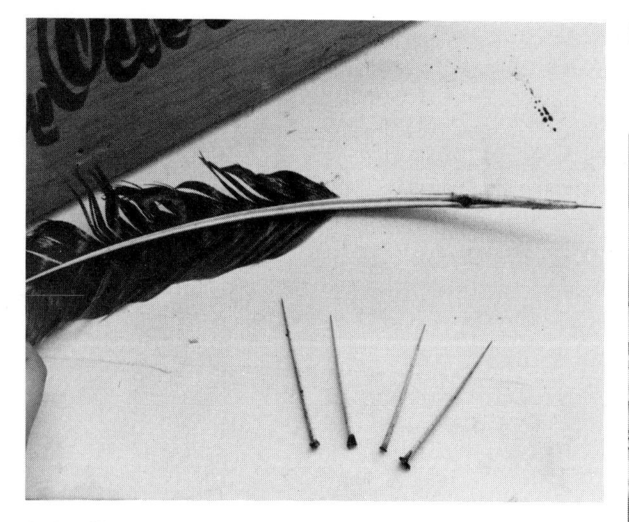

A clay-filled lacquer is coated over another color of unfilled lacquer in a process the Mexicans call *laca rayada.* These thorns, attached to a quill, are used to scratch away patterns.

Lines and shapes are scratched into lacquer by Eduardo Ayala Jimenez of Mexico.

Two *laca rayada* boxes. One is painted with oil paints in the panels.

In another process, clay is not added to the lacquer but, rather, many coats are applied until the buildup is as thick as an eggshell. Then, areas are carved or scratched in the lacquer with a metal pointed instrument. Powdered color and insect lacquer are mixed to form a paste that is rubbed into the spaces with fingers. Only a thin layer of this lacquer fill is applied at any one time. Each application is permitted to dry fully before more lacquer is rubbed into the cavities. The process is repeated until the cavities have been filled and feel like an even coating when fingers pass over the surface.

If colors are desired within the filled area, cavities may be scratched or carved and filled as previously. This process can take months because of the time it takes for the lacquer fillings to harden.

Procedure for Decorating with Lacquer

1. Prepare wood surface: sand until smooth, seal, coat
2. Apply first layer of lacquer. Allow to dry
3. Apply dry clay-mixed lacquer; or, in the second process, many additional coats
4. Scratch away areas not wanted—either positive or negative spaces
5. In second process, fill, via fingers, with lacquer mix until level with original surface
6. Allow lacquer to dry thoroughly, until hard and scratchproof

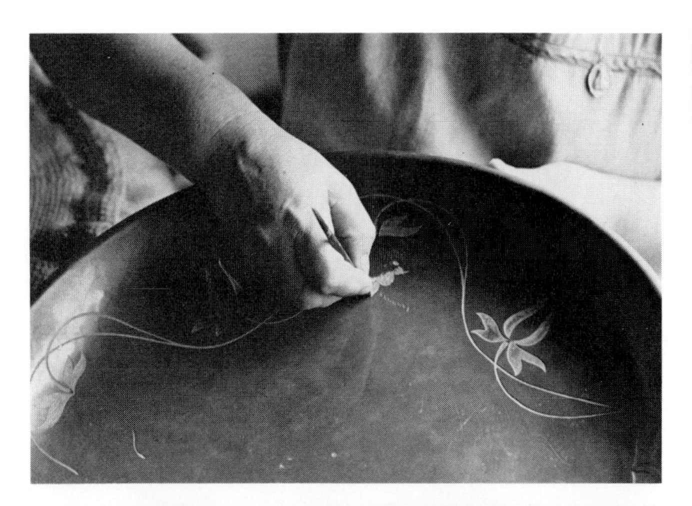

In another lacquer process, lacquer is inlaid, or filled with lacquer. First, a wooden bowl is being sanded. After many coatings, areas within the lacquer are scratched away.

The lacquer, which is suspended in a resinlike paste, is applied to the surface . . .

. . . and rubbed into the cavity in many thin layers. Each layer is allowed to dry thoroughly before the next is applied, and each color is added separately.

An example of this ancient technique that dates back to the original Mexicans, with a design that is of Spanish influence.

Lacquer painted papier-mâché box from Iran.

Peggy Kent applied gold leaf in a crackle effect over lacquer.

The box is lined with Indian silk and with a beaded fabric trim along the walls.

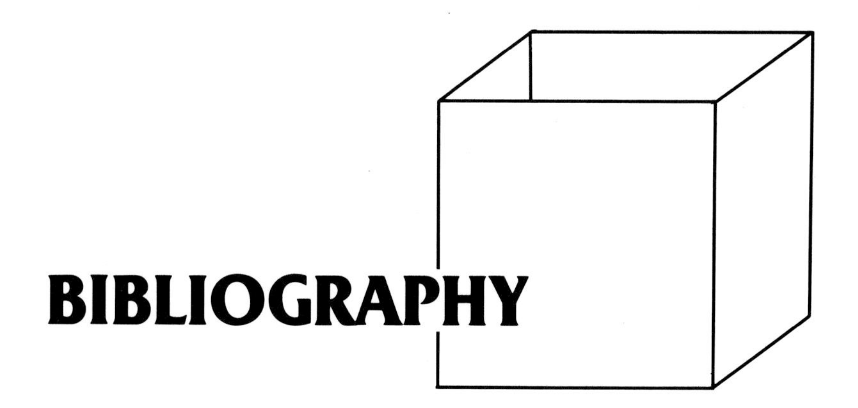

BIBLIOGRAPHY

Bates, Kenneth F. *Enameling Principles.* New York: The World Publishing Company, 1951.

Held, Shirley E. *Weaving: A Handbook for Fiber Craftsmen.* New York: Holt, Rinehart and Winston, 1973.

Lechtzin, Stanley. "Electrofabrication of Metals." *Craft Horizons,* November, December 1964.

Lowenkeim, Frederick. *Modern Electroplating.* New York: John Wiley and Sons, 1963.

Meilach, Dona Z. *A Modern Approach to Basketry.* New York: Crown Publishers, 1974.

Newman, Jay Hartley and Lee Scott. *Plastics for the Craftsman.* New York: Crown Publishers, 1972.

———. *Wire Art.* New York: Crown Publishers, 1975.

Newman, Thelma R. *Contemporary Decoupage.* New York: Crown Publishers, 1972.

———. *Crafting with Plastics.* Radnor, Pennsylvania: Chilton Books, 1976.

———. *Leather as Art and Craft.* New York: Crown Publishers, 1973.

———. *Plastics as an Art Form.* Radnor, Pennsylvania: Chilton Books, 1969.

———. *Plastics as Design Form.* Radnor, Pennsylvania: Chilton Books, 1972.

———. *Quilting, Patchwork, Appliqué, and Trapunto.* New York: Crown Publishers, 1974.

———. *Woodcraft.* Radnor, Pennsylvania: Chilton Books, 1976.

Newman, Thelma R., Jay Hartley, and Lee Scott. *Paper as Art and Craft.* New York: Crown Publishers, 1973.

Nilaūsen, Barbara. *The Process of Electroforming and Its Application to Ceramics.* Iowa City: University of Iowa, 1972.

Rossbach, Ed. *Baskets as Art.* New York: Van Nostrand, Reinhold Co., 1973.

Spiro, Peter. *Electroforming.* London: Robert Draper, Ltd., 1971.

Untracht, Oppi. *Metal Techniques for Craftsmen.* New York: Doubleday and Co., 1968.

Znamierowski, Nell. *Step by Step Weaving.* New York: Golden Press, 1967.

INDEX

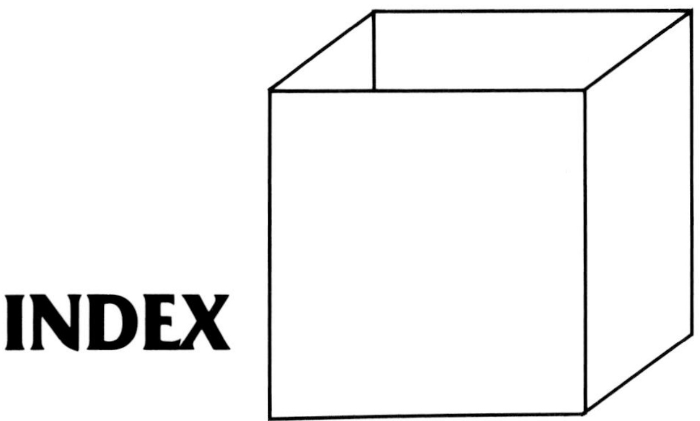

Pages in italics refer to illustrations.